Appropriating Technology

Acting with Technology

Bonnie Nardi, Victor Kaptelinin, and Kirsten Foot, editors

Appropriating Technology

How We Make Digital Tools Our Own

Pierre Tchounikine

The MIT Press
Cambridge, Massachusetts
London, England

The MIT Press
Massachusetts Institute of Technology
77 Massachusetts Avenue, Cambridge, MA 02139
mitpress.mit.edu

The MIT Press would like to thank the anonymous peer reviewers who provided comments on drafts of this book. The generous work of academic experts is essential for establishing the authority and quality of our publications. We acknowledge with gratitude the contributions of these otherwise uncredited readers.

This book was set in Stone Serif and Stone Sans by Westchester Publishing Services. Printed and bound in the United States of America.

Library of Congress Cataloging-in-Publication Data

Names: Tchounikine, Pierre, 1964– author
Title: Appropriating technology : how we make digital tools our own /
    Pierre Tchounikine.
Description: Cambridge, Massachusetts : The MIT Press, [2025] | Series:
    Acting with technology | Includes bibliographical references and index.
Identifiers: LCCN 2025002140 (print) | LCCN 2025002141 (ebook) |
    ISBN 9780262553872 paperback | ISBN 9780262385572 epub |
    ISBN 9780262385589 pdf
Subjects: LCSH: Human-computer interaction | Technology—Social aspects
Classification: LCC QA76.9.H85 T395 2025 (print) | LCC QA76.9.H85 (ebook) |
    DDC 004.01/9—dc23/eng/20250520
LC record available at https://lccn.loc.gov/2025002140
LC ebook record available at https://lccn.loc.gov/2025002141

10 9 8 7 6 5 4 3 2 1

EU Authorised Representative: Easy Access System Europe, Mustamäe tee 50, 10621 Tallinn, Estonia | Email: gpsr.requests@easproject.com

Contents

Preface

Appropriation is the process by which technologies become instruments for us—that is, basic means, basic resources that we use without any conscious explicit effort in the course of our personal and professional activities. As we need to write a text, we transparently open our word processor. As we need to communicate with colleagues or relatives, we transparently open our email account or a communication app such as WhatsApp. As we want to share something with friends, we transparently open one social network app or another, such as Facebook or Instagram. And so on. These applications have become means via which we work, communicate, and socialize.

As I will show in this book, appropriation is a constructive and developmental process. This understanding raises a number of important questions. Is the output—that is, our way of using the technology—coherent with our desires, needs, and interests? To what extent do we literally adapt ourselves (and not just our activities) to technologies? What forces play a role in this process? Are we happy with the current technical offer only because we are unaware of how far we adapt to technologies and accept the social and/or designers' views that come with them? To what extent do we believe we use technologies as we want just because we are unaware of the drivers of our uses? And can designers design *for* appropriation—that is, in a way that increases the probability that users will appropriate the designed technology?

This book addresses these questions, and many others, by proposing a holistic understanding of the appropriation phenomenon. Its contents include an argumentation of why it is important to understand appropriation and to address it as such; an analysis of the phenomena at play; a holistic activity-centered account; how the way we develop (we evolve, we learn) and the underlying psychological mechanisms play a fundamental role; how our human high-level needs also play an important—though often unconscious—role; how the current adaptation means offered by most technologies give us some degree of freedom to align technologies to our needs and desires, although this is far from being fully satisfactory; and, finally, how these design aspects may be improved. Appropriation is a phenomenon stemming from people, and it is impossible to define a theory or a model predicting if and how a technology will be appropriated. Nevertheless, designers have a responsibility in how technologies mediate people's activities. Rather than proposing yet another design approach, I will put forward a certain number of points requiring attention and propose notions, principles, and strategies to frame design *for* and *from* appropriation.

A specific characteristic of this book is that it addresses and articulates the three key dimensions of appropriation processes: *psychological mechanisms*, such as the role of our perception and meaning-making mechanisms, how we develop ways of using technologies, or the role of high-level needs related to our human nature (e.g., self-esteem or engagement); *social forces*, including how culture, workplaces, or collectives influence our perceptions and activities and, crucially, how they convey socially shaped ways of doing things and of using technologies; and *the technological features and properties*, such as their adaptation mechanisms. Why and how we appropriate technologies develops from the interplay of the forces, constraints, and possibilities that arise from these psychological, social, and technical dimensions.

The perspective I propose thus argues against different naive and/or arguable statements such as "Designers can design the human-computer interaction" (when the human-computer interaction emerges); "Iterative user-centered design cycles allow designers to understand and meet users' needs" (when people and thus needs evolve, regardless of whether the design is good or bad); "We use technologies as we decide to" (which ignores the fact that technologies and their associated discourses convey socially shaped ways of doing, how crystallized psychological constructions shape our perceptions and uses, or the fact that we are often engaged in unconscious activities that impact our uses of technologies); "Offering adaptation means to users is useless because they usually do not use them" (which ignores the evidence that users do use adaptation means, albeit those that are often neglected by designers); "Why and how people appropriate technologies can be addressed at the level of abstract users such as 'young people,' or collectives such as work teams" (actually, although the social and collective dimensions are indeed important, the central consideration should be the individuals' characteristics, motivations, or needs); or "Designers should focus on offering usable features and adaptations means only, and users will do what they want with them" (which is simply refusing to solve the problem of producing designs that both empower and respect people).

Basic users of everyday technologies (and actors interested in emancipation concerns) may be specifically interested by gaining an understanding of why and how we use digital technologies, how it affects our activities and ourselves, and how technologies increase our agency but also constrain it. Researchers may be specifically interested in the proposed conceptualization of appropriation, the theoretical advance, the synthesis of the different works (social sciences, psychology, computer science) that address or are relevant to the appropriation topic and, finally, the proposals for the evolution of the field. Designers may be specifically interested in the means (notions, principles, and guidelines) for analyzing the causes of inefficient appropriations, overcoming obstacles to usage evolution (i.e., why people do not adopt technological possibilities that functionally fulfill their needs), and offering relevant adaptation means. Teachers and students may be specifically interested in the material (notions, statements, theoretical elements, examples, extensive bibliography) to include/use in masters and PhD courses acknowledging that, in the human-computer interaction field, the human comes first.

The arguments I develop in this book are based on (1) theories that help make sense of the psychological and social phenomena underlying appropriation and (2) empirical studies of the uses of everyday technologies that have been published in the international scientific literature. With the delay separating the rise of a new technology and the publication of empirical studies of its uses, the time at which I wrote this book (2020–2024) led me to mainly take illustrations from email (the way this "old" technology is interestingly still extensively used is indeed linked to the appropriation phenomena), social networks such as Facebook, communication apps such as WhatsApp, word processors such as Microsoft Word and LibreOffice Writer, collaborative platforms such as Slack, automation platforms such as IFTTT (If This Then That), and, of course, smartphones. There is no doubt that in a few years, these specific applications and/or these types of technologies will have disappeared, or, a minima, evolved, and that new ones will have appeared. However, the mechanisms and forces I present in this book will still stand and will impact how these new technologies are appropriated, because these principles come from us, the users, and our constructive human nature. Actually, although digital technologies do present important specificities, the understanding of appropriation proposed in this book can, to a large extent, be applied to nondigital artifacts.

Acknowledgments

This book builds on a number of other works. I had the privilege of interacting with the authors of some of these works before and/or during the writing of this book. I would particularly like to thank Pierre Rabardel, especially for our discussions about the psychological instrumental genesis theory and the notion of instrumental direction; Victor Kaptelinin, especially for our exchanges about the notion of functional organs and, more generally, activity theory; and Noam Tractinsky, for our discussions about the technology acceptation model tradition. Clay Spinuzzi provided very useful advice regarding the genre approach and the individual/collective articulation. I also benefited from interesting discussions with many colleagues, including Liam Bannon, Michel Beaudouin-Lafon, Susanne Bødker, Wendy Mackay, Carolyn Miller (selfies/not-selfies example), Emmanuel Sander (analogies), and Gérard Vergnaud (schemes). The way I build from or refer to their works is, of course, my responsibility alone.

The series editors Bonnie Nardi and Victor Kaptelinin provided continuous support and advice.

Joanna, thank you for the language editing, and also for having noticed how our use of the "comment" feature of the word editor went beyond exchanging about the task in hand to wider discussions of our respective personalities: your comment made me confident that the book would indeed help readers to reflect on their practices and how they appropriate the technologies they use.

This work was made possible through the support of my institution: Université Grenoble Alpes, Laboratoire d'Informatique de Grenoble (France).

1 Introduction

Do you use email?
. . .
How many emails do you have in your inbox?
. . .
Does this include emails that have already been read?
. . .
Why do these emails stay in your inbox?
. . .
OK, and do you send emails to yourself?
. . .

Conducting our personal and professional activities leads us to mobilize a variety of means that increasingly include digital technologies: We exchange with colleagues or relatives via communication applications such as email, instant messaging apps, collaborative platforms, and social networks; we write texts via word processors; we manage our individual or collective to-do lists via electronic agendas and other specific apps; we close the shutters via dedicated smart-home buttons; and so on.

The very first time we encountered these applications and devices, they were just technical proposals that we were offered for use. What has happened since?

Of course, very basically, we have learned how to use the technology. We have understood its features and mastered the technical skills required, if any.

However, more fundamentally, we have *appropriated* the technology: We have associated it with some of our professional or everyday tasks or goals; we have developed our way of using it in order to achieve these tasks, a way that may be coherent with how the designers expected us to proceed, or one that may prove to be more idiosyncratic; in some cases, we have adapted the technology to our needs or ways of doing; and when we are confronted with the tasks for which it has become a resource, the technology is now immediately and transparently (without any conscious explicit effort) thought of and used.

Typically, the very first time we had a smartphone in our hand, we explored how to switch it on or change the ringtone. In other words, we considered the technology. We now take out our smartphone to ask a colleague something, inform friends, check our bank account, or simply for entertainment—that is, we use it to conduct our activities. Moreover, we do this transparently, in the same way that some of us put our glasses on when we decide to read a text. Checking out what is going on (the task) has become synonymous with picking up the smartphone (this way of doing is immediately and transparently thought of and implemented), contacting a colleague or a relative has become synonymous with opening email or WhatsApp, exchanging with friends has become synonymous with going on one social network or another, and so on.

Two important points become evident when we consider the gap between our first encounter with the technology and the moment where it has become an instrument for us—that is, an actual means, a basic resource for one or several tasks.

The first point is that appropriation is a constructive and developmental process. Since the first time we held a smartphone or considered a new application on our laptop, neither of these artifacts have changed. What has changed is *us*: We have developed our own ways of addressing the tasks we consider and our ways of using these technologies in this context—that is, we have developed the psychological constructions that underlie these behaviors.

The second point is that appropriation fundamentally stems from our activities. Although we sometimes explicitly attempt to learn how to use a new technology, we do not decide to appropriate a technology. What happens is that while we address our professional or everyday activities, we appropriate the means we use, which progressively become basic resources

Let's illustrate this point with a basic task such as remembering information or things we have to do. We have all developed our own ways of addressing this task with the help of artifacts such as taking a pencil and jotting the information down on a piece of paper, using a note-taking or calendar app on the smartphone, taking a picture of whatever must be remembered asking the voice-based smart virtual assistant to add the information to the online to-do list, or writing the information as an email and sending it to ourself to be sure we will come back to it later. According to the strategies we have adopted, we have appropriated the means that were involved. This includes extending our previous use of the technology (e.g., we now use the smartphone camera or the email for a new task) or using a new technology (e.g., starting to use the calendar app or voice assistant), which illustrates well that appropriation is different in nature to the initial process of learning to use a technology. Indeed, many people learn how to use technologies such as the calendar app or voice-based smart virtual assistants in the context, and for the purpose of, remembering things they have to do. However, in direct contrast, it is *because* they already master and use the email application, and constantly have an eye on their emails, that some people turn their inbox into a to-do list. Mastering the technology is just a condition of appropriation. As a matter of fact, we all have some applications on our computers and smartphones that we know about, are able to use and could benefit from, but that we do not use.

Another illustration is that we all develop our own ways of socializing, which leads some of us to adopt and appropriate different applications in different—and rapidly evolving—ways (e.g., WhatsApp for sharing information quickly with friends and family, Facebook or Instagram for disseminating content, and TikTok for entertainment), and to ignore others. Similarly, if we focus on devices rather than applications, it may be said that most of us have appropriated the smartphone for a variety of tasks: phoning, checking the time or the weather forecast, taking pictures, remembering things to be done (via one application or another), socializing, and other tasks.

Let's analyze the constructive nature of appropriation in closer detail. Consider the appropriation of smartphones to know what time it is. What is at play here is very basic: the clock application, the smartphone design (when we move the device, it immediately wakes from sleep and displays, among other things, the clock information), and our general behavior (we always have the smartphone with us). However, let's now consider how the social life of young people involves technologies such as WhatsApp, Facebook, or Instagram. Studies into how young people use these different communication apps (e.g., Nouwens et al., 2017, or Boczkowski et al., 2018) show that what is key is the complex communication system that individuals and groups have built across these applications by attributing them specific purposes and/or connotations (e.g., collaborating with colleagues, exchanging with relatives, having fun with friends or sharing efforts with a community) and using the applications accordingly. In other words, the means mediating the socializing activity is an output of what these users have constructed via their practices, and not simply an output of the technology as such. As an unexpected use, keeping emails to remember the things to be done is another good illustration of this point: What supports this practice is the way we see the inbox as a to-do list and, importantly, manage it accordingly (see section 1.2).

Appropriation is thus an output of the fact that we are active, rather than passive, recipients of technologies, and that in some sense we create our own instruments (our means, our resources) out of the technological substratum that we are offered.

If we consider the big picture, appropriation is a general and pervasive phenomenon. Early humans appropriated natural objects (e.g., stones or sticks) and later created, appropriated, and iteratively improved artifacts. Digital technologies are not exceptions to this rule.

However, understanding why and how we appropriate digital technologies has become a specific challenge—if for nothing else then because it is now practically impossible to avoid their use. For all of us, everyday life may include digitally mediated activities such as making an appointment or even consulting the doctor online, managing administrative tasks via office applications or dedicated websites, booking a table at a restaurant, or shopping, some of which may be difficult to accomplish without using these technologies. Similarly, communication apps and social networks have become an important tool for social life, if not the most important for many people. Also, many workers have to use both dedicated applications such as accounting systems or enterprise resource planning systems and more general technologies including word processors, spreadsheets, or communication applications (e.g., email or collaborative platforms such as Slack). It is thus important to understand why and how we appropriate these digital technologies (or not), identify the impact of this appropriation on our activities and on ourselves, and help designers to design *for* appropriation. I will develop these arguments in section 1.1.

Moreover, the appropriation of digital technologies presents specific characteristics. As outlined above, the way social forces shape the appropriation process is core. How these technologies may be adapted also plays an important role, as we will see later in this book. Another specificity is the time span. Consider the set of artifacts used in a professional activity such as blacksmithing or those used for cooking for one's family. These artifacts evolved over very long periods and were often passed from one generation to the next with little or no changes, along with the associated ways of addressing tasks. In direct contrast, the electronic devices we use today significantly differ from those we used a few years ago and from those we will use in few years. The applications we use on these devices develop faster still. We must thus constantly discover, use, and (possibly) appropriate new applications and even new devices (such as cell phones, smartphones, smart home controllers, smart speakers, activity trackers, 3D printers). And, importantly, we have little control over this technological progression: The operating system of our computer may not look the same tomorrow after its automatic upgrade, and some of the applications we use now will not work on the computer we will buy in a few years.

What we have seen in this first section can be summarized as follows: The appropriation of digital technologies is a natural phenomenon; it develops from, and is constitutive of, how we actually use things in practice; it has a strong impact on us because our specific appropriations of these technologies (i.e., the ways in which we appropriate them) mediate and thus shape—or hinder—our activities; and these technologies have become central to our professional and personal lives.

On the basis of this first general overview, let's now look at the following in more detail:

- why it is important to understand appropriation (section 1.1),
- why it is necessary to consider appropriation as a phenomenon in its own right (section 1.2), and
- why a new perspective on appropriation is needed (section 1.3).

This will be the basis for presenting

- the perspective adopted in this book (section 1.4) and
- the structure of the book (section 1.5).

1.1 Why Is It Important to Understand Appropriation?

Understanding why and how we appropriate technologies is important for both (1) the users, as individual and groups, and (2) the designers and promoters of these technologies.

Users' Benefits: Emancipation and Agency

The impact of digital technologies on our lives is huge. Understanding how and why we appropriate technologies thus provides us with more control and agency over our lives, and is emancipatory. Let's list a number of important points.

Understanding why and how we adapt to technologies   As we will study in detail, the way we appropriate digital technologies shapes the way we use them, which in turn impacts our activities and thus also shapes, to some extent, what we are. Since pioneering works such as Mackay (1990), it has been shown that appropriation is a coadaptive phenomenon: People both adapt to technology and adapt this technology to themselves.

The first point I want to make is straightforward: If we adapt to technology—that is, adapt our activities and our psychological structures—as indeed we do, we should at least be aware of this and understand the underlying phenomena.

Deciphering our uses   The Socratic dialogue used as an epigraph at the outset of this chapter will have made many readers realize that, although they had perhaps never thought about it, they do actually use emails to remind themselves to do things, or, in other words, they have appropriated their inbox as a to-do list. As another example, many people use social networks to share photos or experiences. However, a closer examination of the real motivations underlying such uses may reveal the user's need to fulfill high-level requirements such as presenting the self or socially reinforcing his/her worldview.

What surfaces here is that questioning our appropriation of technology improves our awareness of what we use technologies for—that is, the actual activities that we carry out. Raising such an awareness is a sine qua non condition for a reasoned and controlled use of technologies. As we will see in chapter 5, many appropriations are driven by essentially unconscious activities or needs. Although fully understanding what one does and why would probably be an endless and intractable undertaking, gaining a certain degree of awareness is both beneficial and emancipatory.

Analyzing and improving our practices   One of the benefits of better understanding how we use technologies and why we develop these uses is the opportunity to inform reflective analyses of the efficiency of our practices and, possibly, their evolution. Consider the example of using the email inbox as a to-do list and consequent practices such as sending emails to oneself. The design rationale of email applications is to communicate with others, not with oneself. As individuals, we have the right to use applications for other concerns than those initially considered by their creators. Even when this may not be the most efficient means to address the task at hand, this appropriation remains positive from an agency perspective, insofar that it is a conscious decision (see preceding point) and that we are happy with the output. Typically, in this case, a reflective analysis may lead users to take action to improve this strategy (for example, through the use of folders and filters to keep the inbox size manageable and make to-do emails visible) or, in direct contrast, change the practice (for example, by using dedicated task manager features or applications).

As another illustration, let's consider communication apps. Empirical studies reveal how people often face issues related to how the technology properties fit the diversity of interactions for which they are used. For instance, the way in which some applications prompt awareness information, such as when a user was last connected, may be in line with how young people interact with their friends, but it may be problematic for managing their interactions with their family. This can be understood by considering the appropriation of these technologies as idiosyncratic communication places with their own specific membership rules, purposes, and/or emotional connotations (Nouwens et al., 2017). Understanding this kind of phenomena may pertinently inform the selection, customization, and/or uses of these apps, and thus improve our control and agency.

Being aware of the role played by rules, norms, and accumulated experience
When we use and appropriate technologies, we actually also appropriate the socially accumulated experience they reflect, which influences both our behaviors and our mental functioning (Kaptelinin & Nardi, 2006). Let's explore this important point further.

Wertsch (1998), who studied mediated action in the light of Vygotsky's perspective, nicely highlights the fact that "we are unreflective, if not ignorant, consumers [of cultural tools]" (p. 29) with the example of how we solve a problem such as multiplying 343 by 822: We align the numbers of intermediate multiplications—that is, use a culturally shaped means (in his words, the solving of the problem is accomplished by "I and the cultural tool I employed"). As a tangible example, a corkscrew or, more precisely, the device plus the knowledge of how it should be used, reflects the experience of other people who tried to solve the "opening bottles" problem.

The fact that, when creating digital technologies, developers embed not only some means to accomplish tasks but also some rules and norms (Orlikowski, 1992) is very obvious in applications such as accounting systems, workflows, and enterprise resource planning systems. However, this is also the case for basic technologies. For instance, the email features and associated knowledge reflect a culturally shaped way of exchanging letters and documents, which largely stems from the accumulated experience of postal email and organization management (see for example the notions of "correspondents," "CC," or "Attached documents"). Although we could to a certain extent decide to act differently from the commonly accepted usual way of doing, it actually shapes most usages. Similarly, word processors reflect a culturally shaped conceptualization of what a written document is, and now come with explicit templates such as "inventories," "newsletters," "resumes," or "cover letters." Likewise, the way social networks encourage "liking," "sharing," or communicating about oneself also reflects cultural constructions.

As an example of analysis specifically regarding this type of phenomena, Spinuzzi (2003) studied the activity of traffic safety workers over thirty years. He shows how every new generation of technologies (from paper maps to GPS databases) actually reflects and carries socially shaped ways of doing that are both stable and evolutive. I will come back to how these social constructions impact our perception and meaning-making mechanisms in section 2.4.

By appropriating technologies, we thus also appropriate the rules, norms, and/or experiences that they convey, and even reinforce them. It may be useful to be aware of this, consider what these norms are, how we want to act, and be particularly aware that technologies may primarily serve the interests of other entities.

It may be thought that the provision of more or less free access to a wide range of options (e.g., dozens of different communication or social networking apps) allows us to act as we decide to because we can choose the applications that meet our needs and desires. This is indeed the case for some people. However, as we will see in chapter 2, many people do not decide to use the technologies they use, which are more or less imposed on them by explicit constraints (e.g., in workplaces) and/or by social forces (e.g., using the same communication means as their relatives or friends). Moreover, it may also be worth considering the extent to which we believe we use technologies as we want just because we are unaware of the drivers of our uses. Do young people make social networking applications their own, or does the contrary occur?

Being aware of the economic aspects Finally, an important specificity of digital technologies such as communication apps or social networks is data production, and this data has an economic value. Ekbia and Nardi (2017) have labeled "heteromation" the "computer-mediated mechanism of extraction of economic value from various forms of human labor through an inclusionary logic, active engagement, and invisible control" (p. 39). They highlight how sharing material (e.g., photos or texts) or commenting about it (e.g., likes) via communication platforms actually produces social data such as reusable resources or individual profiles. Tech companies benefit from the economic value of this data without paying anything in return. Actually, the platforms are intentionally designed to create a specific type of user who not only communicates but who shares—that is, a user who produces the data that generates income.

Although one may argue that we obtain compensation in the form of practical benefits (e.g., expressing ourselves or keeping in contact with others), I would say that, whatever our personal opinion may be about this deal and our definition of the future it may lead to, it is important and emancipatory to be aware and understand what is happening. One aspect is the understanding of the economical mechanisms—that is, modern capitalism functioning (see Ekbia and Nardi's analyses). Another is why and how we appropriate and use these technologies as individuals. This includes conscious but, also, unconscious motivations (see chapter 5); specific cognitive mechanisms (chapter 4); and the affordances of the technologies used (chapter 2).

From Users' Interests to Designers' Duties

The point made above is that it is important for us as individuals and groups to be aware of the mechanisms underlying our uses of technologies. This understanding is emancipatory and can inform the decisions we take such as selecting one technology rather than another or adapting the used technologies to our needs or desires.

Nevertheless, now that possessing and using digital technologies has become commonplace, if not mandatory, we, as individuals, also have the right to be given the opportunity to benefit from the positive outputs of appropriation, and to avoid the use of technologies that hinder our activities.

The next section explores the designers' interests to understand appropriation in this light, which drives the overall perspective adopted in this book: The focus is not on design considerations that aim to facilitate appropriation as opposed to nonuse, but rather on enhancing the likelihood that users will appropriate the technology in a way that fulfills their needs, desires or values—that is, the designs respect the users rather than control or constrain them.

Designers' Benefits: Designing for Appropriation

Designing for appropriation In addition to being a use phenomenon, appropriation may be considered a desirable goal. The fact that technologies are appropriated by their users is of interest to the companies designing these technologies, to the companies or institutions who use these technologies to conduct their business or mission, and, as mentioned above, to the users.

For companies designing technologies, the fact that these technologies are appropriated and used increases their benefits, whether directly (when the technology or the service is sold) or indirectly (via advertisements and users' data). As well illustrated by social networks or office tools, the way a small number of technologies become the mainstream means of addressing some tasks creates a social force that acts as a lever of expansion. Adoption and use, however, may stem from the fact that the technologies allow people to conduct their activities as they wish or result from commercial strategies such as putting pressure on the organizations buying these technologies (e.g., by imposing a specific office suite) or on the users (e.g., by rendering the use of other office suites difficult by using proprietary formats, or by designing social networks features that trigger dopamine, FOMO—fear of missing out—or addictive behaviors). The former, which is indeed preferable, requires to help designers design *for* appropriation.

For companies or institutions who use technologies to conduct their business or mission, the rationale for using technologies and introducing new means is generally to support and/or change their employees' practices. The improvement of work efficiency is the prototypical motivation, but other goals such as the preservation and development of workers' wellness and health can be cited. In such cases, it makes sense to enhance the probability that people will appropriate the technology in a way that is coherent with the rationale for its introduction, and/or deal with the difficulties users face due to the negative outputs of any unexpected appropriations. What is important here is not whether people have adopted a technology, but rather the characteristics of the uses stemming from the different factors and forces involved.

Finally, the fact that users benefit from resources that facilitate their activities (communicating with friends, managing documents, remembering things to be done, and other activities) is indeed positive. When the initial studies of software appropriation were developed in the 90s, they logically focused on workplaces—that is, where these technologies were mostly used—and considered how taking the factors supporting/hindering appropriation into account could improve workers' efficiency. Now that possessing and using digital technologies has become commonplace, this efficiency concern must be extended to basic users and personal tasks or life events.

The fact that appropriation is an emerging process, and is thus not entirely predictable, does not mean that designers cannot be supported in designing for appropriation—that is, enhancing the chances that appropriation will develop. As we will see, a better understanding of the mechanisms at play is a helpful conceptual resource for a variety of tasks such as analyzing the causes of inefficient or problematic appropriations; identifying and addressing obstacles to evolution—that is, why people do not adopt technological possibilities that technically fulfill their (apparent) functional needs; informing the design of releases; improving applications interfaces; or offering users pertinent adaptation means.

The cornerstone of these tasks is to make sense of the uses that develop, and why they develop. Let's develop this point.

Making sense of uses The fact that people sometimes use artifacts and technologies for unexpected tasks—that is, for purposes that were not initially anticipated by designers—is striking and highly illustrative of the appropriation phenomenon. Nondigital examples include using a stone as a hammer, a pencil to fix a hair bun, or an oven to warm up a room. The use of email as a to-do list or for archiving data is a classic example. Many other cases of function creeps can be cited—for example, using smartphone cameras as mirrors or spreadsheets as presentation applications (Dourish, 2017).

In addition to their illustrative nature, odd appropriations have the specific interest of highlighting the possible role of unanticipated forces and/or factors. This is useful for deciphering the appropriation phenomenon.

These very specific cases, however, should not distract us from the big picture: The most frequent pattern is that applications are used for the purpose for which they were initially designed. Indeed, it is precisely on these very occasions that an in-depth understanding of appropriation may prove to be specifically useful. In such cases, it is much more difficult to identify, decipher, and, when necessary, respond to the impact of appropriation processes on uses.

A good illustration is the use of collective work platforms such as Slack (2020). These applications are designed to allow companies or groups to create a global chatroom that can be broken into different specific channels such as task-related, topic-related or subgroup-related channels. Creating a specific set of channels and then systematically posting messages in the appropriate ones presents core advantages such as solely contacting the relevant persons, providing group awareness, or constituting dedicated archives. However, such platforms may also be used to communicate in more traditional ways such as sending emails or one-to-one messages, which is what happens in certain cases. For instance, Stray and Moe (2020) report an empirical study of the use of Slack in an international software company that has offices in different countries, leading to cultural differences between users. The authors mention that while workers who were used to open team communication wanted the discussions to be visible via channels, the developers who were accustomed to more one-to-one communication and hierarchical organization tended to continue sending personal messages.

Considering such cases to be non-appropriation is misleading. What happens is just that some users appropriate and use the new technology in a way that does not correspond to expectations and that is actually more similar to how they appropriated previous technologies. Similarly, some people only use their smartphones to make calls: They have appropriated the device . . . as a basic phone.

However, companies or institutions that hope to gain efficiency from the expected uses need to make sense of the real use users make of these tools. The interest of a theoretical account of appropriation such as the one presented in this book is to draw attention to the different (social or psychological) forces and/or factors (e.g., the technological properties) that may play a role. (I will use *factor* as a generic term.)

In the Slack example mentioned above, what is at stake does not seem to be a design issue. It is not a matter of a failure in requirement analysis, a missing or inappropriate technical feature, or a poor user interface. The explanation should rather be sought in how people often import patterned ways of doing things from one setting/technology to another. (Hereafter I will simply refer to *ways of doing* or *ways of using*, meaning patterned ways of addressing tasks, patterned ways of using technologies.) This phenomenon will be studied in detail in chapter 4. It calls for a closer study of why individuals appropriated the preceding technologies in this way, and identifying what prevents the practice from evolving. For instance, some workers remain aware of the hierarchical organization of the workplace, and the implications thereof, and act accordingly, whether consciously or unconsciously. As another illustration of this point, the way people appropriate collaborative editing tools and the awareness features they propose (e.g., comments, track-changes, or revision history) is affected by how these users address the expected task (collaborative editing) and other tasks such as managing social relations or self-esteem (see chapter 5).

As one can see from these examples, what is at stake here is not, as methods such as GOMS (goals, operators, methods and selection rules) (John & Kieras, 1996) propose, to define the set of actions that is needed to achieve a particular goal with the system and predict the interaction time of an ideal skilled user. What is at stake is to understand *why* particular practices develop.

It could be argued that the different psychological or social factors that led to a particular appropriation could, or indeed should, have been anticipated during the design phase. This is more easily said than done, and it is also misleading. It is simply impossible to fully predict what will happen, because it will be the result of something that did not exist before the technology was used, namely the actual activity of the users.

1.2 Why Is It Necessary to Consider Appropriation as a Phenomenon in Its Own Right?

One of the main ideas underlying this book is that appropriation is a complex phenomenon—that is, a phenomenon that involves different factors interacting in a systemic way. Attempting to understand appropriation from one perspective only, such as its social or technical dimensions, is thus necessarily reductive and oversimplistic. Appropriation must be addressed as such—that is, by considering it as a phenomenon in its own right and identifying the different forces and factors that may be involved.

Any detailed study of the uses of a given technology highlights how the specific appropriations developed by people stem from the interplay of different factors. I will take as an illustration the findings for the "appropriation of email as a personal information management device" case. The specific interest of this example is that it has been studied in dozens of empirical analyses, conducted over several decades, which have provided strong evidence for the multiplicity of factors involved in appropriation (Tchounikine, 2019a). As this book is not focused on email, I will simply mention the aspects of these works that relate to appropriation and enhance this example with illustrations from other technologies.

Impact of Social Practices
Since the 2000s, analysts have pinpointed that email has become more like a habitat than an application, where many people spend a lot of time (Ducheneaut & Bellotti, 2001). This is still the case for many workers. Other individuals now "live" in their social networks or digital games.

The fact that these applications become habitats, which is an output of social forces such as work practices or socialization, has effects per se. For those who always have an eye on their inbox, keeping the emails that refer to things to be done is indeed an efficient way to remember them. Moreover, this process reinforces itself. A survey in a large tech corporation revealed that most users sent emails to themselves, some of them doing so once or several times per day (Bota et al., 2017). Similarly, the fact that some young people spent most of their time on online games led them to turn the VoIP or instant messaging applications of these games into their main communication means (which, in turn, led the technologies to evolve; see the example of the Discord platform). The same occurred for Facebook and, later on, for other social networks. As we will see, the uses of Facebook, WhatsApp, or Instagram are shaped by social forces (how other users use them) more than by their technical specificities.

These examples highlight how work or social practices and, more generally, the institutional structures within which people act, are core to the appropriation phenomena.

Development of Idiosyncratic (Low-Level) Practices
The study of how people appropriate their email has revealed general behaviors. This includes communication strategies—for example, preference

toward email communication in scenarios raising interpersonal risks or face threat (Joinson, 2004)—and unexpected appropriations for task management (e.g., providing reminders for current tasks, tracking task status, or maintaining information relevant to those tasks), personal archiving, or contact management (Whittaker et al., 2006; Haraty et al., 2016).

However, these works have also highlighted that what happens at the level of the individual is often highly idiosyncratic (Whittaker et al., 2006; Bellotti et al., 2005; Ducheneaut et al., 2001). For instance, studies reveal that some of the users using their emails as reminders stick to the basic strategy of keeping all emails reminding them of something to be done in the inbox. Others deal with constant pollution by incoming emails by creating a "to-do" folder where they file the emails, thus creating a to-do list. In direct contrast, some users file all their emails in subject-related folders except the to-do emails, which makes the latter clearly visible in the inbox. Another (nonexclusive) strategy is to create a "to-do" tag and use it as a filter. And these are just a few examples. As mentioned by Bellotti et al. (2005), "It is easy for the analyst to get bogged down studying the infinitely variable (and often fascinating) practices of e-mail users" (p. 102). Similarly, as we will see, the widespread use of smartphones has led to general practices such as exchanging selfies, but here again idiosyncratic group practices develop.

Role of Design and Technical Properties

How email application interfaces make the inbox salient and visualize the emails as lists is arguably of central importance for using emails as reminders, and the strategies described briefly above are based on the filing, tagging, filtering, or sorting features of applications. In other words, email applications provide technological affordances (a notion that we will study in detail) for the development of such strategies.

In recent years, another technical factor impacting users' strategies is the multiplicity of devices used, which may include laptops, desktops, tablets, smartphones, and/or smartwatches. For instance, an email that was opened early in the morning on a smartphone may no longer appear as a new message and, thus, will not be highlighted when the same user opens his or her workplace computer email application later on. This leads some users to develop different workarounds depending on the device (Cecchinato et al., 2016). Strategies to manage this issue include changing the status of the email to "unread" (many applications offer this arguably useful but also odd feature: Why would anyone say "Well, I read this email, but let's say I didn't"?) or forwarding the message to oneself. General technical factors such as the adopted or imposed technical protocol (e.g., if messages are kept on the server and mirrored on the application or downloaded to the device and removed from the server) also allow, suggest, or prevent email management strategies.

The technical characteristics of applications do therefore have a strong impact on their appropriations. If email clients only presented emails one by one as in early applications or did not offer the means (e.g., filing, sorting or tagging) that allow the idiosyncratic strategies mentioned above, they would not have been appropriated in the same way and/or for the same purposes. Similarly, although social forces affect the communication or social applications we use (it makes little sense to use communication apps that our friends or relatives do not use), the individual characteristics of these applications (e.g., picture-editing tools or vanishing messages) also play a role.

Role of User Fundamental Needs

The works conducted in the 2000s shed some light on why people use email to remember things to be done: Emails are often related to tasks that must be achieved; conversely, tasks to achieve are often referred to in emails; the email application is constantly open and checked; and email clients offer filing, tagging, filtering, or sorting features.

However, more fundamentally, the central cause for such uses to develop is our need for to-do lists. This need, of course, existed long before computer technology, and we had already developed other means to address it such as paper Post-its. Emails became another way to achieve this goal when computers and online communication became a part of our everyday life. In a similar way, the use of social networks to share photos or experiences stems from our need to present our identity and to socially reinforce our worldview, both of which are tasks that we previously addressed—and continue to address—with other means such as dress codes or participation in clubs.

The point made here is that before being users, we are humans. Our fundamental needs or desires as humans influence our appropriation—or non-appropriation—of technology (and, as I have already argued, the contrary is also true: Our appropriation of technology influences our needs or wills as humans). As a way to emphasize this point, let me return to the Socratic dialogue used as an epigraph for this chapter. I adapt this dialogue slightly in public presentations, as follows: "Do you use email? Yes? OK, here is a series of questions. *Please answer, but in your head.*" The reason for this is that many people feel uncomfortable with some of the answers or, rather, what their colleagues or friends may think of these answers.

Appropriation is related to what we do, and what we do is related to many things including what we are and/or want to be (although, indeed, our acts are sometimes opposed to our wishes or hopes). Characterizing how people use technologies and deciphering how and why they made them their own may lead to very personal and possibly sensitive topics. It is very obvious that the way people appropriate and use social networks applications is impacted by fundamental needs such as self-esteem or being connected to other individuals. As we will see, this is also the case for how some individuals appropriate and use their email application or their word processor.

Psychological Dimensions

When we have appropriated a technology and turned it into a personal instrument, we immediately and transparently access this resource and use it This is the case for the corkscrew we use for opening a bottle of wine, our glasses to read a text, WhatsApp for sharing news with relatives, or the email to oneself if this has become the basic way of remembering things to be done.

This state of affairs has a psychological substratum: We have developed the set of psychological constructions that implements (1) the task-technology link (e.g., the link between remembering things to be done and sending an email to oneself); (2) the fact that the technology is immediately thought of when we are confronted with the task; and (3) the way of achieving the task and the way of using the technology in this context.

What surfaces here is that people develop (in the sense of developmental psychology that is mentioned above) and that appropriation is part and parcel of this development. Actually, as I will argue, development is the fundamental mechanism of appropriation.

As users, we may thus take decisions such as "I will now use this technology for this task." However, technology can only become a basic resource that we transparently access and use if we develop specific psychological constructions. The decision to use a technology does indeed favor this development, but it is only a preliminary step. For instance, many users take the decision to use electronic agendas means but, in practice, continue to use means such as emails or word documents (Haraty et al., 2016). In the words of

Carroll et al. (2002), they *adopt* the technology, but then do not *appropriate* it and often finally abandon it.

The developmental processes underlying appropriation may be subconscious, which is well illustrated by the email example. People are often surprised to realize that they actually turned their email inbox into a to-do list but were completely unaware of doing so. As a personal example, it's when I started to reflect on appropriation processes that I realized that, when working on long documents with my word processor, I used an undo-redo sequence to take the cursor back to the last edited paragraph. Prior to this, and although it had been part of my basic practice for many years, I never considered this sequence of action as a means to an end: It was a method (a way of doing things) that had emerged and crystallized, and that I was using transparently, without considering it as such.

The fact that such developments may be subconscious is not without significance. It means that when we are more or less forced to use a technology by work practices or social pressure, we are likely to develop and integrate ways of doing without being aware of these developments (see the emancipation concerns in section 1.1).

Another important point is that these psychological developments not only impact if and how people appropriate new technologies; they also impact how things may evolve. For instance, email is an old technology that was developed in the 1970s and became part of everyday life in the 1990s. One could have thought that it would be replaced by instant messaging, social networking applications, or, for workers, generic (e.g., Slack) or corporate collaboration platforms. Yet, email practices persist. One of the reasons is that, for psychological reasons in particular, uses change in an evolutionary rather than revolutionary way (see chapter 5). How users import their usual ways of doing things into new technologies—for example, by using collaboration platforms such as Slack as a one-to-one communication channel in a similar way to email—is but one aspect of this general phenomenon.

As well illustrated by in-depth studies such as those conducted on email technology, why and how we appropriate technologies are thus a direct result of the interplay of the forces, constraints, and possibilities that arise from psychological, social, and technical dimensions. Appropriation must therefore be studied as a complex phenomenon—that is, by considering it as such and by studying the different forces and factors that are involved.

1.3   Why Is a New Perspective on Appropriation Needed?

Since the late 1980s a certain number of works, actually surprisingly few, have considered the appropriation of digital technologies as a specific topic. I will present their contributions, throughout the book, when addressing topics for which they provide some useful light. Here, I will focus on why a new perspective is needed—that is, why the existing knowledge on appropriation provided by previous works is insufficient—and identify the issues that need to be addressed. For this purpose, I will consider the following topics: the definition of appropriation, the theoretical background, the focus (collective/individual), and the perspective to design.

Definition of Appropriation

Current state of affairs   Different definitions have been proposed to date.

Referring to Marx, Poole and de Sanctis (1989) highlight that "to appropriate an object [is] to use it constructively, to incorporate it into one's life, for better or worse" (p. 150).

Carroll et al. (2002) define the process of technology appropriation as "the way that users evaluate and adopt, adapt and integrate a technology into their everyday lives" (p. 58). Appropriation is how users transform "Technology as Designed" into "Technologies in Use": "Over time, the technology is stabilised and becomes an integral part of users' activities; we call this appropriation" (Carroll, 2004, p. 340).

Dourish's (2003) definition is probably the most widely cited: "Appropriation is the process by which people adopt and adapt technologies, fitting them into their working practices. . . . [It] might involve customisation in the traditional sense (that is, the explicit reconfiguration of the technology in order to suit local needs), but it might also simply involve making use of the technology for purposes beyond those for which it was originally designed, or to serve new ends" (pp. 465–467).

According to Dix (2007), appropriation is when "the technology has been domesticated," "the users understand and are comfortable enough with the technology to use it in their own ways," and "the technology has become the users' own not simply what the designer gave to them" (p. 28).

Schwartz et al. (2015) mention that "appropriation represents the tactics of everyday practice, which give artefacts their individual meaning and results in use, which might be both unforeseeable and unintended" (p. 555).

Required improvement   Although the definitions listed above all make sense, draw attention to some relevant points and resonate with the analyses presented so far, they address appropriation as an umbrella construct—that is, a "broad concept or idea used loosely to encompass and account for a set of diverse phenomena" (Hirsch & Levin, 1999, p. 200). They sketch the phenomenon rather than propose a precise definition, and they mix notions (e.g., adoption, integration in practices, or adaptation of technologies) that differ in nature.

As raised by Hirsch and Levin, umbrella constructs are sufficient for intermediation purposes such as sharing areas of interest. Moreover, they may have pragmatic advantages in some cases: Broad descriptions may best capture the inherent complexity and messiness of the empirical world (Suddaby, 2010), while adopting one precise definition and a "validity police" perspective may entail limits and risks (Hirsch & Levin, 1999).

Nevertheless, the vagueness of umbrella constructs also impedes understanding and knowledge capitalization. The notion of *usability* is a good example: Its usual definition ("the extent to which a system, product or service can be used by specified users to achieve specified goals with effectiveness, efficiency and satisfaction in a specified context of use"; ISO 9241–11:2018) is multidimensional—that is, mixes different aspects. As argued by Tractinsky (2018), this contributes to the fact that the usability umbrella construct is instrumental for some design practices but is of little use in *understanding* the interaction between humans and computers.

In direct contrast, a precise and instrumental definition of appropriation is needed to address the emancipation and design considerations listed in section 1.1. We need a definition that builds on an understanding of the phenomenon (and not simply a description of the outputs), which dissociates the intrinsic (necessary) aspects of appropriation and the contingent aspects such as symptoms or possible implications, and which allows the identification of criteria.

Theoretical Background

Current state of affairs   While some works make general references to phenomenology (e.g., Gallagher, 2014), current appropriation-oriented analyses of technology uses have mostly built on one of three general research traditions, which I list here in no specific order.

The first is the *structuration perspective*, which studies how social systems are created and reproduced through social structures (Giddens, 1984). This social perspective has been the general frame and/or the starting point of different structural-oriented analyses of technology appropriation (Poole & de Sanctis, 1989; de Sanctis & Poole, 1994; Orlikowski, 1992, 2000; Vyas et al., 2017).

The second is *activity theory* (Kaptelinin & Nardi, 2006), which suggests focusing on users' activities and considering appropriation from an expansive learning perspective (see, e.g., Stevens and Pipek, 2018). As we will see, the activity theory also proposes an understanding for different psychological aspects of importance when considering appropriation, including mediation, perception of affordances, and the notion of functional organs. Although the instrumental genesis approach proposed by Rabardel (2001, 2003) originally stems from the ergonomics research tradition, it comes close to, and complements, the activity theory perspective.

The third is the *genre perspective*, which studies how typified rhetorical actions based in recurrent situations become routinized (Miller, 1984, 2015). Although initially focusing on texts, this approach provides general insights on how artifacts mediate activity and, of core importance, how technologies carry socially constructed ways of doing (Russell, 2009; Spinuzzi, 2003).

Required improvement   Although these general theoretical approaches do propose relevant explanatory contributions that I will build on, appropriation is a complex phenomenon involving social, psychological, and technical dimensions (see section 1.2). Building on just one of these theories, like in previous works, thus addresses only part of the different dimensions of appropriation only; it is reductive and oversimplistic.

As an example, the study of the email case in the light of the structuration and genre approaches did revealed important points such as the key roles played by typified actions, traditions of addressing tasks, and traditions of using artifacts. For instance, Yates et al. (1999) identified a dozen explicit or implicit genres structuring how the members of a firm involved in a particular project communicated via email—for example, official announcement, trip report, release notice, or team report. Johnson et al.'s (2012) analysis of email networks in a company showed that gender, tenure, or hierarchical boundaries played a weaker role in email network than in offline networks. More recently, Sevtsuk et al.'s (2022) analysis revealed how email uses may be affected by the spatial structure of the workplace.

However, such analyses do not consider other dimensions such as the cognitive (developmental psychology) aspects or, more generally, the user as an individual, and the technological factors are solely considered from the structural perspective rather than in and of themselves. They do not allow us to understand how transparent uses of email develop, which requires the consideration of mediation and psychological mechanisms. They also fail to explain why we may feel anxious (or proud) to reveal to others (and sometimes even to ourselves) how and why we manage our emails, why we use emails to remember things, or the specific role of technical characteristics such as the structure of the inbox or the filing, tagging, and filtering means. Conversely, the works that focus on the characteristics of technologies tend to ignore the social or psychological dimensions.

Such shortcuts are problematic for users and also for designers, who cannot design for appropriation if they are unaware of some of the appropriation drivers. Typically, understanding the social dimension is key, but it is not instrumental if one is not aware of how psychological forces may facilitate or hinder user evolution. The definitions listed above are symptomatic of this issue: They make relevant points about some aspects of the phenomenon but also lead to confusions (e.g., consider "adoption" or "technical mastering" as synonyms of "appropriation").

The improvement that is needed here is to acknowledge the complexity of the appropriation phenomenon and to change the analytical approach. Rather than considering appropriation *from* how one general theory (e.g., the structuration perspective, activity theory, or the genre perspective), we should identify the different factors that may play a role and then articulate the different theories that shed some light on these factors. Adopting this perspective is, for instance, what made me discover that high-level activities such as defending one's worldview or the psychological mechanisms related to the "extended self" (see chapter 5) sometimes play a nonanecdotal role.

Another positive aspect of such a perspective is to provide a framework for future works. I have little doubt that the overall conceptualization of appropriation proposed in this book and, in particular, the review and articulation of factors that may play a role, can and will be enhanced in the future by the identification of additional factors and/or dynamics.

Focus (Collective/Individual)

Current state of affairs   The works that have considered appropriation so far have emphasized its collective and organizational dimensions. This is the cause for, or output of, the adopted theoretical backgrounds: These dimensions were indeed put forward by the structuration, genre, and activity theory perspectives (and more precisely, the "third generation" of the latter, as developed by Engeström, 1987, 2001, 2009).

Required improvement   Although the collective and organizational dimensions are core to the appropriation phenomena and it would make no sense to ignore them, the individual and idiosyncratic dimensions must be given more importance, and more centrality, than in the works to date. I will support this claim with two arguments, which differ in nature: first, the psychological dimension of appropriation, and second, the evolution of the technical ecosystem.

The first reason why a focus on individuals is crucial is the developmental dimension of appropriation, which has not (with the exception of Rabardel's work) been sufficiently considered to date. As already mentioned, the state of affairs where the technology *mediates* our activities and we use it *transparently* (I will come back to these notions in chapter 2) does not occur by magic. We have developed the psychological constructions (namely, we have adapted our previous mental schemes and/or created new ones) that drive our perception and understanding of both the activity and the technology and, in turn, allow us to make the connection and shape our acts, and we have developed *ways of doing* activities and *ways of using* technologies (see chapters 2 and 4). These developments are individual, although, as I will explain in chapter 4, they have an inherent social dimension (we develop via our interactions with others). This includes interactions with the traces of others in the technologies we use—that is, the socially accumulated experience that the devices reflect and the associated ways of doing things. Development is both individual/idiosyncratic and collective.

The second reason for the necessity of a focus on individuals is the increasing empowerment of individuals by the technical ecosystem, and the consequent implications in terms of emancipatory concerns (see section 1.1). Let me develop this point in detail.

In the early decades (1960s, 1970s, and 1980s), digital technologies were thought of as something that is created by designers and proposed to users. While computers were essentially used in the work practice, this was just continuing an existing pattern: The tools are offered to or imposed on workers, who have to use them, and must thus adapt their practice accordingly. Structural dimensions were key. In the following decades (1990s, 2000s), computers started to appear in our everyday life. The offer evolved, and users were provided with some elements of choice, such as being able to choose the application they bought or used (albeit from a somewhat limited range) or adapting a few aspects of applications via a "preferences" menu. Nevertheless, the general picture remained similar.

What has changed in the recent period is that we have gained more agency—that is, more capacity to act in our current environment. We have much more latitude in deciding the technologies or services we use. Installing a new application on one's computer or smartphone or enhancing existing applications via plug-ins is commonplace, even for basic users. The application and plug-in offer is wide and often free (or, rather, as mentioned above, paid for with our personal data or our consent for advertisement; as we will see, this may play a role in non-appropriation). Customizable Internet-of-things (IoT) buttons may be associated by the user with one or several features accessible via the network: for example, smart home devices that manage a thermostat, turn on the light, or open the automatic rolling shutter. When using these technologies, users literally define what the device is a means for. Automation platforms such as IFTTT (2020) allow us to combine technologies with each other and, for example, to state that if a message is received in a given application then a specific action should be triggered in another. And so on.

Nevertheless, the real agency gains are far from being satisfactory, and one of the reasons for this is another recent evolution: As individuals, we have to some extent internalized that digital technologies are essentially smart means that are conceived and designed by private tech companies, then generously proposed to us. This is very different from user-centered design (UCD) and particularly participatory design (PD) frames of mind. Far from being a means for improving system usability and user satisfaction alone, PD was originally a way to address agency in the professional realm and to engage workers and their organizations in the development of technologies as part of this agenda: "Participatory design sought . . . to intervene in situations of conflict through developing more democratic processes" (Bannon et al., 2018, p. 27). Later on, UCD and PD addressed agency in everyday life. However, as highlighted by these authors, this dimension seems to have vanished. What happens now is that social forces more or less impose on us the use of different technologies in our everyday family and social life (and in many working situations)—for example, office tools, communication applications, or social networks. In turn, this leads us (or actually more or less obliges us) to appropriate them and deal with all the consequences, including the promotion of the underlying values of these technologies or, on the cognitive side, our internalization of certain ways of doing things. However, while we saw in section 1.2 how we tend (and indeed need) to develop idiosyncratic uses, the technologies we are offered are essentially generic innovations based on a "one design fits all users" (and all uses) design. And these technologies come with an implicit "if you don't like it, don't use it" message that is not without issues.

The improvement that is needed here is an understanding of how individuals appropriate technologies as individual actors and not solely as members of a collective. This enhancement is needed to (1) help users understand their appropriations and practices and be in control of their uses and (2) help designers to do their duty—that is, offer technological proposals that allow individuals to benefit from the positive outputs of appropriation and avoid the use of technologies that hinder their activities.

Design

Current state of affairs   A pertinent general descriptive life-cycle model of technological innovations appropriation has been proposed by J. Carroll (Carroll et al., 2001; Carroll, 2004). The baseline is the design of a new application or device that opens up new possibilities. This technological proposal may not retain the user's interest—that is, it will become a not-adopted technology (but may be reviewed for one reason or another at a later date). When users are interested and adopt the technology, they start using it. This may lead them to take possession of the technology in a way that satisfies their needs (appropriation). Finally, a technology may be appropriated but, later on, no longer satisfy the users' needs (which the authors refer to as disappropriation).

Within this general perspective, managing the uncertainties related to how people will appropriate digital technologies may be framed within J. M. Carroll's well-known task-artifact cycle (Carroll et al., 1991). This classical conceptual model features the coevolution of tasks and artifacts. (Carroll would later mention that *activity*—what people actually do—conveyed what he meant better than *task*.) In his words, "Human activities implicitly articulate needs, preferences and design visions. Artifacts are designed in response, but inevitably do more than merely respond. Through the course of their adoption and appropriation, new designs provide new possibilities for action and interaction. Ultimately, this activity articulates further human needs, preferences, and design visions" (Carroll, 2014). This may be addressed via classical UCD or PD approaches.

Several researchers (e.g., Beaudouin-Lafon, 2004) have pertinently argued that the analyses should focus on user interaction rather than technical interfaces, a position that is now widely accepted. This has led to a shift in focus from designing interfaces to "designing interaction," which has become a label.

Finally, some general design guidelines have been proposed. In a seminal article, Dix (2007) argues for providing elements where users can add their own meanings, exposing design intentions, or supporting use rather than controlling it. Pipek and Wulf (2009) argue that designers should consider "the entirety of devices, tools, technologies, standards, conventions, and protocols on which the individual worker or the collective rely to carry out the tasks and achieve the goals assigned to them" (p. 455). I will discuss these works further in chapter 7.

Required improvements   Although these works provide useful general guidelines, there is room for improvement and complements. As we will see in chapters 6 and 7, the understanding of appropriation proposed in this book and, in particular, the consideration of psychological phenomena (e.g., the role of users' conceptualizations) allows the definition of more precise and instrumental guidelines.

However, more fundamentally, taking an appropriation perspective suggests—and indeed imposes—the consideration of design in a way whose premises fundamentally differ from the currently accepted and applied *interaction design* approach.

According to the Interaction Design Foundation website (Siang, 2020), interaction design "is the design of the interaction between users and products" (digital products like apps or websites) and "the goal of interaction design is to create products that enable the user to achieve their objective(s) in the best way possible." I can only agree with such a goal. However, if primarily focusing on the design of interfaces is indeed a dead end, shifting the focus to the design of interaction is, although rhetorically interesting, an oxymoron.

An appropriation perspective highlights that designers do *not* design the interaction between users and products: This interaction emerges and fundamentally stems from the users, their actual activities, and the idiosyncratic and situated nature thereof. In other words, designers do not have any control over this interaction. They have an impact and this impact may be core, but it remains limited to the technical conditions of the user-product interactions. As we saw in the preceding sections, the personal resource that develops via the appropriation process may significantly differ from its original technical substratum. For example, people's social life is mediated by the complex communication system they have built across communication and social networking apps by attributing them specific purposes (rather than by the technical characteristics of these applications), and remembering things to be done with emails stems from how people see the inbox as a to-do list and manage it accordingly. We will see other examples throughout the book.

Spinuzzi (2003) already warned us against the "designer-as-hero" and "user-as-victim" trope, which may lead to consider that "[workers'] innovations and feedback are useful only when designers consolidate and shape them to support the work models that the users are not able to understand" (p. 9) and maximizes the role played by commonalities. This has unfortunately remained a very pertinent point. And, in the recent period, the way technologies are presented as proposing "a new experience" reveals a designer-as-magician trope, which is, here again, fairly paternalistic.

Acknowledging that how people use digital technologies in their practices is not fully predictable and, moreover, is evolutive (see chapters 3 and 4) has a clear-cut implication: Design approaches that do not acknowledge and take into account appropriation processes are intrinsically limited. UCD, which pays extensive attention to the characteristics of users, and PD, which involves stakeholders in design and, more generally, in careful requirements analyses, are and remain useful. Nevertheless, these processes take place in the design phase (i.e., prior to actual uses, although tests and iterative cycles may be organized) and address users in the plural form (i.e., an abstract entity, rather than actual actors with their idiosyncratic characteristics, motivations, and activities).

The improvement that is needed here is not the replacement of the methods or techniques used in interaction design, UCD, or PD. Rather, the needed improvement is to engage designers in a conceptual move and help them prepare for action through a conceptual framework: Appropriation must be considered as a phenomenon rather than a goal; this phenomenon and the forces playing a role must be understood and acknowledged; design must be addressed as creating, in this light, the technical conditions of the psychological developments underlying appropriation; and the latter objective can be addressed in the light of precise design guidelines (see chapter 7 and synthesis in the conclusion chapter).

Now that we have seen why it is important to understand appropriation, to consider appropriation as such, and to develop a new perspective, let's introduce the specific perspective adopted in this book and how it will be explored.

1.4   Perspective Adopted in This Book

Points Made So Far

Understanding why and how we appropriate digital technologies is of importance for both users and designers. What is at play for users is a set of emancipation and agency concerns: understand why and how they adapt to technologies; decipher their uses; analyze and improve their practices; be aware of how they also appropriate norms and accumulated experience; and be aware of the economic aspects. What is at play for designers is to make sense of uses and inform design for appropriation, which ties in with the users' interests to benefit from the positive outputs of appropriation and avoid technologies that hinder their activities.

A number of important points have been introduced, which I will study in more detail throughout the book: Appropriation is a constructive process that stems from us and our activities rather than from technology alone; it is constitutive of how we actually use things in practice and is not simply learning how to use a technology; it involves psychological mechanisms (for instance, our appropriations and uses are related to our perception and meaning-making mechanisms, and when we appropriate technologies we develop, i.e., we adapt our ways of doing things and/or elaborate new ones), social forces (for example, culture, workplaces, or collectives convey socially shaped ways of doing things and of using technologies), and, last but not least, technological features and properties, particularly their adaptation mechanisms.

Finally, I have listed a number of needs: a precise and instrumental definition of appropriation; the identification of the different factors that may play a role and an articulation of the different theories that shed some light on these factors; closer attention to individuals, their idiosyncratic behaviors and their psychological characteristics and developments; principles and guidelines helping designers to (1) engage in a conceptual move and consider appropriation as a core phenomenon, (2) understand why and how appropriation develops, and (3) create the technical conditions for this development.

General Line of Thinking

The perspective adopted in this book focuses on individual's activities; articulates the psychological, social, and technological dimensions of appropriation; and considers emancipation concerns. The principles and rationale underlying these three characteristics are as follows.

An activity-centric perspective   In this book, I approach appropriation in terms of how users find a way to address the tasks they consider. This is, very explicitly, an activity-centric perspective. The answer to the question, "Why does appropriation develop?" is: "Because we are humans and, as humans, we search for, and develop, resources to mediate our activities."

The implication is that, when considering appropriation, the core question is not "Is the device or application appropriated?" but, rather, "For what activities is it appropriated?" A device or application is appropriated when it has been incorporated into activities and has become an instrument—that is, a usual way of addressing one or several tasks. This is why I use the plural in some places: One user may develop different appropriations of the technology for different tasks, and different users may develop different appropriations of one technology for the same task. Understanding the appropriation of digital technologies calls first and foremost for the identification and analysis of what these tasks correspond to, and how/why the uses of these technologies developed by people allow them to accomplish these tasks.

As we will see in more detail in chapter 4, the user I consider here is thus an individual who acts in and on the world, which is the productive dimension of the activity, and in doing so, changes him/herself and thus develops, which is the constructive dimension of the activity. This focus on activity— that is, what is done and how—leads us to specifically focus on users as pragmatic actors (actors who do) rather than epistemic actors (actors who know). In other words, what drives users is not the understanding of the applications or devices they use but the functions thereof for the user. When we add an appointment in our online calendar, type a text with our word processor, or send an email to ourselves to remember a task, the technical aspects that are at play are the how-to knowledge (the "knowledge in acts") that we have developed and not what we know or do not know (our explicit knowledge) about our smartphone, word processor, or email client.

Considering that appropriation fundamentally stems from users' activities is not, of course, a naive claim that people have absolute agency and that their wills are the unique explanation and driving force of how they use technologies. Similarly, focusing on people as actors does not mean that people are reducible to their acts (see chapter 5).

A focus on the individual   In line with the consideration of users as actors, I focus on individuals—that is, the agent who engages in actions. Typically, I consider why a given individual uses his/her email to remember to-do lists, or how he/she appropriates a given communication app in a specific way, and not how email has changed company practices, how young people engage in social networks, or how a given service changes its organization. As mentioned in section 1.3, the rationale for this focus includes how the technical ecosystem increasingly empowers individuals and, also, how people develop.

Importantly, taking individuals as the analytical entry point is not to be confused with restricting analysis to individuals: Activity is always social (this will be discussed further in section 2.6). We have already seen the role of work practice, and it is very obvious that the adoption and use of communication apps by young people is influenced, if not shaped, by social forces, which may supersede individual dissatisfaction with an incumbent service (Meier et al., 2021; Sun et al., 2017). The approach developed in this book acknowledges these forces but addresses them as impacting forces. Incidentally, although this is not central to the proposed approach, we will see that the actor in some cases is a group, for example when workers decide collectively about how they will use an application.

An emancipatory perspective   Wright (2010) argued that an emancipatory social science that seeks to generate scientific knowledge relevant to the collective project of challenging various forms of human oppression and creating the conditions for human flourishing requires (1) the elaboration of a systematic diagnosis and criticism of the world as it exists; (2) the consideration of viable alternatives; and (3) an understanding of the obstacles, possibilities, and dilemmas of transformation.

Although this book is not structured in this way, I share this general analysis. This perspective shapes the analyses developed in this book in different ways.

A first implication is to consider the appropriations and uses of technologies that people develop as such, and not in relation to other stakeholders' perspectives or interests. People's uses of technologies can be considered as expected/unexpected or efficient/inefficient for a given task or goal. Alternatively, as proposed by de Sanctis and Poole (1994), their uses can be considered as faithful/unfaithful (i.e., consistent/inconsistent with the spirit and structural feature design). They should not, however, be considered as good/bad. It is particularly important to dissociate the fact that technologies are used as expected from the fact that the actual use is dysfunctional or problematic, which is a different question.

A second implication is to consider what Wright labels as "oppression," an arguably strong term that I would contextually redefine as hidden impacts, controlling influences, or unnecessary technical constraints hindering idiosyncratic uses of technologies. In line with this perspective, taking how people develop resources for themselves as an analytical entry point is not simply a means to explicitly acknowledge the evolution of the technological offer. It is also a means to address the users' needs and rights mentioned in section 1.1.

Thirdly, as using digital technologies has become intrinsic to many human activities, the obstacles to uses must be identified and addressed. Although using these technologies does not necessarily make us better or happier people, the difficulty or impossibility of using them may reduce our overall agency and/or create personal issues (see chapter 5).

Finally, digital technologies have a specific characteristic: They can be rendered highly adaptable. Design considerations may thus include avoiding "oppression" but, also, open up opportunities for user adaptations. There is room to realistically sketch out a more attractive future than the present state of affairs (see chapter 7).

Methodological Approach

Addressing a phenomenon in the light of different theories, and in a way that recognizes both its individual and collective nature, requires a leading analysis axis. This section presents the rationale for the articulation of the different theories and research traditions adopted in this book and then the way it refers to existing empirical studies.

General analysis axis   What drives the appropriation of a technology is its usefulness and, more precisely, the fact it mediates some of our activities. Understanding why and how appropriation develops thus requires understanding what makes us perceive a technology as useful, why and how it becomes a mediator (or fails to do so), and the characteristics of this mediation.

The psychological phenomena involved here (e.g., perception, mediation, meaning-making, or development of ways of doing things and using objects) are not specific to digital technologies. They stem from our human nature, which did not change on the emergence of these technologies. Similarly, the involved social forces (e.g., structures or genres) are general forces. However, the way these phenomena and forces apply is specific to what is at play (i.e., addressing activities with digital technologies), to the characteristics of these technologies, and to the specific interplay that this context creates. As already argued, this is why appropriation must be addressed as such.

What is to be elaborated is thus not a new stand-alone general theory but, rather, a holistic account of appropriation articulating the findings of the different theories and research traditions that have studied these phenomena and forces.

The theoretical account of appropriation I propose in this book builds on findings from the technology acceptance model research tradition, activity theory, the instrumental genesis theory, the structuration approach, the genre approach, psychology works by authors explaining how we develop concepts or skills, the ecology of artifacts and places approaches, and works studying some aspects of our human nature such as the self-determination theory or experimental existential psychology. Although the rationale for mobilizing these frameworks will appear more clearly as the detailed argumentation develops throughout the different chapters, the following sections present a preliminary picture of this theoretical landscape.

Users' needs and motivations   Within the technology acceptance model research tradition, thousands of empirical works have produced coherent evidence of the role of *perceived usefulness* in the adoption of technologies and, thus, in their uses. These works provide a pertinent entry point for considering the topic of technology usefulness.

However, appropriation is not just adoption. When considering appropriation, the notion of perceived usefulness must be addressed in a way that differs significantly from the adoption perspective. Activity theory (first and second generation) sheds a very relevant light on this topic and, in particular, allows us to make sense of the specific appropriation-motives link.

Developmental aspects As I have already argued, development (i.e., the fact that people change, evolve, learn) is the core mechanism of appropriation. When appropriating a technology, we modify our previous cognitive resources and/or develop new ones. The factors playing a role here include the concepts we use to make sense of our activities and of the technologies we use, and the psychological constructions related to our patterned ways of doing and ways of using (namely our mental schemes). It is thus necessary to go further than the general statement asserting that appropriation is a coadaptation process, which only implicitly acknowledges that users' interactions with artifacts lead them to evolve. In the absence of an understanding of this evolution phenomenon, it is impossible to address the implications.

In addition to a general understanding of how activities are driven by motives, activity theory provides the conceptual means to address the mediation, perception, and meaning-making mechanisms underlying these developmental aspects. Vygotsky's works on mediation and learning, and other cognitive works addressing the notion of mental scheme, provide additional insights. Finally, the instrumental genesis theory, which stems from ergonomics, provides a specific understanding of the development of instruments. Although different, these research traditions share many common ideas and notions and are theoretically coherent.

Socially conveyed ways of doing things and using technologies As I have already highlighted, activity, and thus appropriation, are in no way purely individual. We do not act in the middle of nowhere: The structures (e.g., cultures, professions, or institutions) we are part of, and the technologies we use, convey worldviews and ways of doing/using.

Although activity theory also addresses this point, the structuration approach and, more interestingly, the genre approach, shed specifically pertinent light on these phenomena and how they shape appropriation. Here again, these research traditions have some links. Genres may be considered as social structures, with individuals interpreting and enacting genres through communicative practices (Yates & Orlikowski, 2007). Both activity theory (third generation) and the genre approach consider organizations, and these approaches may be theoretically and practically articulated as described by Spinuzzi (2003) and Russell (2009).

Users' needs and motivations (continued)   Users' motives may be multiple (because people are often simultaneously engaged in several activities) and may possibly relate to different levels. Acknowledging this complexity is core.

The fact that people are often simultaneously engaged in several activities, with high-level activities/motives breaking down into lower-level activities/motives, has been addressed by different research traditions that consider workplaces. Staying with the activity theory, a good example is how Engeström (2001, 2009) enhanced the initial theory to study networks of interacting activity systems in organizations (AT third generation). While not specifically devoted to appropriation concerns, the analytical means developed in this work are highly relevant for deciphering the motives-activities links and, more specifically, the collective dimensions.

Other types of high-level motivations, however, have been underexplored and merit some attention: The theories studying human nature have demonstrated that it leads us to address fundamental needs and motivations such as preserving self-esteem, developing social bonds, or acting coherently with our values. These motivations do impact and, in some cases, shape how we appropriate some technologies. To support this claim I will present the evidence provided by theories that have already been used to make sense of uses (e.g., self-determination theory or extended self theory) and then show that further explorations are highly necessary by presenting an original analysis of how the experimental existential psychology (XXP) and the existential philosophical approaches may also help. The objective here is to show that understanding appropriation requires an open-minded holistic perspective and, as I do not claim completeness or perfection, suggest and pave the way for future analyses.

These theories, and particularly those addressing existential concerns, are arguably fairly broad and depict pervasive general forces. In no way can they replace analyses that contextually and historically address organizations or activity systems, as activity theory or the genre approach both suggest and support. In other words, one cannot develop an understanding of the appropriation phenomenon by building on an existential perspective, for example. Nevertheless, high-level human needs generate high-level motivations that sometimes turn into low-level considerations, which is the rationale for considering the theories addressing these needs as complementary means that may potentially provide additional information. For instance, I will show in chapter 5 that some uses of communication apps or word processors are not fully understandable if we do not consider the impact of some of the high-level motivations stemming from our human nature.

Illustrations and empirical evidence   As argued in the preceding sections, appropriation is an emergent phenomenon. It is therefore not a question of the elaboration of a predictive model and its evaluation via a dedicated empirical experiment, but rather the understanding of the processes and forces at play, and an argumentation of this understanding.

The existing theories I refer to (see above) have been studied via specific works and empirical studies. I will therefore focus on how their findings (which I will not discuss as such) shed light on appropriation.

I will support the additional specific claims I make in this book by (1) referring to empirical studies of technologies that have been published in the scientific literature and/or (2) factual technical analyses (what the application allows the user to do, and how). The rationale for referring to several examples and existing empirical studies (rather than one new example and its empirical study) is threefold. First, the forces playing a role in appropriation are multiple and apply differently according to cases. Taking just one example and an ad hoc punctual empirical study would thus only allow me to illustrate and/or support part of my arguments. Second, although appropriation of digital technologies occurs much faster than for many other artifacts, it is nevertheless a matter of weeks or months and, importantly, stems from in-context (and not in-lab) uses, that is, how we use technologies in the course of our basic activities. Referring to works whose rationale is to study basic uses of technologies such as email, social networks, or word processors makes it possible to benefit from the findings of such in-context studies (typically, the rationale for using email as a running example is the stock of empirical studies). Finally, considering different everyday technologies highlights that appropriation is a widespread phenomenon that is not limited to "complex" systems and impacts many of our activities.

1.5   Structure of the Book

This introductory chapter has analyzed in detail why it is important to understand appropriation for both users and designers and why the elaboration of this understanding requires considering appropriation as such. It has also sketched the phenomena at play and the perspective adopted in this book and introduced the used theoretical background.

Chapter 2 continues the analysis by carrying out a close examination of the phenomena at play in light of this theoretical background. I will first show how and why the notion of perceived usefulness must be addressed differently from adoption-oriented works and why users' technical ecosystem must be considered in terms of *ecologies of artifacts* and *places*. Once this general context has been clarified, I will use activity theory to study the implications of considering that technologies mediate our activities, highlight the core notion of *functional organs*, and clarify the notion of *usage transparency*. These analyses will lead us to consider how we, as humans, make the connection between our motives and means—that is, study our perception and meaning-making mechanisms. For this purpose, I will first introduce the notion of affordance and then study how the structuration theory and the genre approach highlight the existence of socially conveyed ways of doing. This chapter also proposes a discussion of the necessity to articulate both the high/low analytical levels and the individual/social concerns, and ends with a synthesis of the different points made.

Building on this conceptual substratum, chapter 3 describes the activity-centric understanding of appropriation that I propose. In a first step, an analysis of how the notion of functional organs resonates with users' perceptions and meaning-making mechanisms will lead to the definition of the notion of *functional value*. We will then study the resulting general conceptualization of the appropriation process that it leads to and identify the topics that require further analysis.

The first of these topics is human development. For readers interested in the cognitive dimensions and their insights for use analyses (and, later in the book, design), chapter 4 proposes an account of the developmental processes at play when people appropriate technologies—that is, why and how our psychological structures evolve. The topics studied here include the fact that developmental aspects must be addressed via macro- and microanalyses, the role of patterned ways of addressing tasks and using technologies, the consideration of these patterns as cognitive structures, and the central role of users' conceptualizations. These topics shed some additional light on appropriation as described in chapter 3.

Chapter 5 then explores how the appropriation of technologies may be impacted by high-level psychological forces and factors such as personal traits and needs (e.g., competence, autonomy, or relatedness), existential concerns, psychological attachments, or engagement. Here again, these analyses shed some additional light on the appropriation phenomenon.

As mentioned by all works, appropriation is a coadaptive phenomenon. After clarification of the human side of this adaptation (psychological mechanisms, potentially impacting forces) in chapters 4 and 5, chapters 6 and 7 address the technological dimension.

Building on the understanding of appropriation proposed in the preceding chapters, chapter 6 offers both users and designers an analysis of if and how current adaptation techniques offer users some useful adaptation means. The considered techniques include means for adapting applications to personal conceptualizations, defining semantic tagging systems, customizing interfaces, adapting/defining specific behaviors (e.g., automating groups of commands or taking advantage of automated learning features), enhancing applications via plug-ins, associating devices with specific actions or, though very specific, end-user programming. For designers, the proposed analysis of these means shows how adaptation means should be thought about when considering appropriation.

Chapter 7 then studies how the specific analyses developed throughout the book may help to improve how adaptation means are currently designed. The considered topics include the importance of addressing inter- and intra-user variability and, for this purpose, design *for* and *from* appropriation; being attentive to conceptual issues; taking a functional value perspective to user information, which may be used to improve the way one draws users' attention to complementary means or to facilitate the sharing of practices; offering under-designed means to engage users in adaptations; and favoring productive analogies. The specific case of the appropriation of technologies based on artificial intelligence is then analyzed in more detail. The chapter ends with some complements from other design perspectives (technology-enhanced activity spaces, semiotics engineering, automated interface adaptations, and the infrastructuring and meta-design approaches)

Finally, chapter 8 concludes this book with the claim that appropriation should be considered a central notion of the human–computer interaction field, a selection of some of the points made throughout the different chapters, and some reflections on the future of appropriation-centered works.

1.6 Appropriation, Appropriations

The object, rationale, and perspective of this book having now been clearly defined, this final section completes the picture by taking a step aside and discussing other meanings or approaches of "appropriation."

Appropriation comes from the Latin ad (to) + propriare (make one's own). It has different related, but significantly different, definitions: "To set apart for or assign to a particular purpose or use"; "to take exclusive possession of"; and "to take or make use of without authority or right" (online Merriam-Webster dictionary). If we instantiate these definitions in the case of technologies, they all make sense. The perspective on appropriation developed in this book comes close to the "assigning to a particular purpose or use" definition. However, using technologies in a way that is unexpected and/or goes against some explicit or implicit rules also comes close to the "making use without authority or right" definition. Similarly, how a few dominant tech companies impose their products is a structural case of "taking exclusive possession," and their domination impacts how people use technologies. Although one may address the subject of appropriation from the last two perspectives, it is not part of the analysis axis of this book. As a side comment, it may be noted that the notion of "application ownership" is very specific. In contrast to other artifacts, an intrinsic characteristic of software is its easy duplication at very little cost and without depriving the original owner. This process, however, may be encouraged, as in the free and/or open-source movement, or discouraged, for legitimate (e.g., paying the designers of the present and future versions) or more questionable reasons.

If we now consider how appropriation and its emancipatory dimension has been studied in other areas, some of the latter resonate with the perspective adopted in this book. Rather than carry out an extensive review, let's look at a few examples.

The notion of appropriation has been used as a key process for explaining human nature. In particular, Ollman (1971) points out that, for Marx, appropriation "is the interaction between man's senses and nature, in which the powers involved use the nature they come into contact with for their own ends" and occurs "through and after perception and orientation"

(p. 86). What has been specifically referred to in different studies of technology appropriation, and is coherent with the perspective adopted in this book, is that for Marx "[appropriation] means to use constructively, to build by incorporating. . . . The individual appropriates the nature he perceives and has become oriented to by making it in some way a part of himself with whatever effect this has on his senses and future orientation" (Ollman, 1971, p. 89). When considering Marx's heritage, the notion of appropriation has of course also taken a specific negative connotation related to the analysis of how, under capitalism, appropriation leads to dispossession and alienation. This, however, is a different point.

Within a different line of thinking, de Certeau (1980) refers to appropriation and "cultural braconnage" (poaching) to highlight that individuals are not passively dispossessed and subjugated to authorities. People invent their everyday lives, and the appropriation and subversion of space are part of how they develop creative resistance to structures.

The notion of appropriation also appears in works studying the analysis of language. In particular, Bakhtin (1981) states that "language is not a neutral medium that passes freely and easily into the private property of the speaker's intentions; it is populated—overpopulated—with the intentions of others" (p. 294). A word "becomes 'one's own' only when the speaker populates it with his own intention, his own accent, when he appropriates the word, adapting it to his own semantic and expressive intention" (p. 293). Actually, as we will see in chapter 2, another aspect of Bakhtin's legacy is core to the analysis of technologies appropriation: the notion of genre.

Finally, appropriation also appears in the arts. Evans (2009) refers to appropriation as stealing images and forms from other makers. Examples include remixes, ready-mades, détournements, and pastiches. A détournement (in arts: a variation of an existing production that takes on a different meaning) pertinently characterizes the innovative uses of technologies. How task automation platforms offer the means (and predefined patterns) to connect or bridge applications or IoT devices with services may be seen as a remix substratum. In arts, appropriation is here again also given specific meanings, for example when aspects of colonial history are involved.

Although such works are inspiring, a common reference to an abstract meaning does not mean that what this meaning corresponds to for one type of consideration and in one specific context may be derived from what it corresponds to in others. Care must thus be taken to differentiate between general inspirations, frames of mind, metaphors, and slippery imports of concept and principles.

As an example, de Certeau explains his point by considering producers as owners of spaces, who develop strategic spatial practices, and consumers as individuals who live in spaces that do not belong to them but, as poachers, develop local tactics to create their own meanings for space. This perspective may be used to conceptualize some aspects of technology appropriation. For instance, as we will see in chapter 2, Dourish (2006) powerfully uses de Certeau's approach to explain an important aspect of computer supported collaborative work, namely the distinction between activity spaces (i.e., the output of design) and activity places (i.e., spaces invested with specific understandings of behavioral appropriateness). In Dourish's words, "Strategic practices are the practices of design, whereas tactical practices are the practices of use" (p. 302). The mere fact that de Certeau's work may be used to make sense of the space/place dichotomy does not however imply that his general perspective and specific conceptual apparatus applies to other aspects of appropriation. Similarly, the way Bakhtin states that language "lies on the borderline between oneself and the other" is inspiring: His work has influenced, for example, Wertsch's (1998) study of mediated action as the tension between active agents and cultural tools, and mediation is indeed an important aspect of technology appropriation. Nevertheless, this does not imply that there are identical mechanisms at play in the appropriation of language and technologies. As a final example, the activity-centric perspective I adopt leads to an acknowledgment of Marx's point that activity is both productive and constructive: "[Man] sets in motion the natural forces which belong to his own body, his arms, legs, head and hands, in order to appropriate the materials of nature in a form adapted to his own needs. Through this movement he acts upon external nature and changes it, and in this way he simultaneously changes his own nature" (Marx, 1976, p. 283). The implication is not that a Marxian analysis should be adopted but, rather, that research on developmental processes can shed some relevant light on the appropriation-related aspects of the constructive part of activity that Marx, and others, have put forward.

2 Phenomena at Play

Perceived usefulness is an obvious driver of appropriation and is thus a good analytical entry point. However, as I will show in section 2.1, this topic must be addressed differently here than it has been in adoption-oriented works. The implications include taking a holistic perspective to the users' technical ecosystem (section 2.2), analyzing what mediation corresponds to in greater detail (section 2.3), and studying how we make the connection between our motives or needs and the technology or, in other words, the phenomena and mechanisms underlying our perception of technologies and the associated meaning-making mechanisms (section 2.4). I will conclude this chapter by returning to the importance of articulating low/high analytical levels and individual/social forces, which was briefly outlined in chapter 1, and propose a synthesis of the points made. These analyses form the background of the understanding of appropriation described in chapter 3.

2.1 The Role of Perceived Usefulness

Perceived Usefulness as Defined in Adoption-Oriented Works

Definition and main findings The notion of *perceived usefulness* (PU) has been introduced and studied in detail by the researchers focusing on technology acceptance. The rationale for these works is crystal clear. Digital technologies have the potential to improve worker performance, user satisfaction, and, from another perspective, tech company profits, but they can only do so if users are willing to accept and use these systems. It is therefore essential to define models that help predict user acceptance and to deal with potential unwillingness.

The general line of thinking underlying these works is that technology acceptance can be explained in terms of the user's internal beliefs, attitudes, and intentions.

In the initial TAM model (technology acceptance model; Davis, 1989), PU is defined as "the degree to which a person believes that using a particular system would enhance his or her job performance." It is measured via questionnaires including items such as "Using [the system] improves my job performance," "reduces the time I spend on unproductive activities," or "makes it easier to do my job." The unified model UTAUT (unified theory of acceptance and use of technology; Venkatesh et al., 2003), which builds on a synthesis of different perspectives (e.g., the theory of reasoned action, the theory of planned behavior, the innovation diffusion theory, or the social cognitive theory), proposes other notions carrying a similar idea: Job-fit ("the extent to which an individual believes that using [a technology] can enhance the performance of his or her job"), long-term consequences ("outcomes that have a pay-off in the future"), relative advantage ("the degree to which an innovation is perceived as being better than its precursor"), or outcome expectations ("the performance-related consequences of the behavior"). UTAUT also acknowledges the role of social influence—that is, "the degree to which an individual perceives that important others believe he or she should use the new system." Specific tailored versions of the UTAUT model were also proposed, introducing elements such as hedonic motivation, price value, or habit (Venkatesh et al., 2012). A synthesis and a review of UTAUT exploitation can be found in Venkatesh et al. (2016).

The empirical analyses showed that *perceived ease of use* (PEOU) also plays a key role in users' intention to use a given technology, albeit to a lesser extent than PU. PEOU is defined in TAM as "the degree to which a person believes that using a particular system would be free of effort." It is measured via questionnaires including items such as "I often become confused when I use [the system]" or "My interaction with [the system] is easy for me to understand."

While the initial TAM models studied work practices and considered "job performance," the notion of PU can be extended to the actual uses of digital technologies by considering the more general notion of "task performance." In the same way that the use of a specific application may contribute to a given job performance, the use of social networks or communication apps by young people contributes to their social-life performance.

In a nutshell: PU and PEOU (which may be an antecedent to PU) act as a determinant of the user's attitude toward using the technology, which acts as a determinant of the intention to use, which in turn acts as a determinant of the actual use. This model has proved its predictive value and is still exploited.

Analysis from an appropriation perspective Usefulness is core to both adoption and appropriation, and all definitions of technology appropriation reflect this very intuitive point. We only appropriate technologies that we use, and adoption is thus a sine qua non condition for appropriation. Moreover, our reasons for adopting technologies likely also play a role in why we appropriate them. Unsurprisingly, a systematic literature review of research works considering why users intend to continue their use of online technologies reveals that the usefulness of the latter is central to this decision (Yan et al., 2021).

However, care must be taken that, in adoption-oriented works, the perceived usefulness notion explicitly carries the idea that what is at stake is how users *foresee* the outputs of using the technology. Although some works or associated discourses are sometimes confusing, the TAM perspective acknowledges that the *actual* usefulness—that is, the usefulness revealed by actual practices—may be different, which is often the case. Along with other works, the meta-analysis conducted by Turner et al. (2010) showed that TAM variables are a much stronger predictor of the behavioral intention to use a technology than the actual use of the latter.

When taking an appropriation perspective, what needs to be understood is *why* people use technologies (rather than simply *if* they use them or intend to use them), and the notion of usefulness must thus be revisited. First, it is important to take into account that people often use technologies that they have not really decided to use. Second, as mentioned above, appropriation relates to the users' actual uses. Lastly, the tasks for which the technology is used may differ from both designers' and users' expectations and, in some cases, may actually be unconscious. Let's consider these different (though interrelated) points.

People Do Not Necessarily Decide to Use the Technologies They Use

Although users have an increasing capacity to decide what applications they want to use, considering that the technologies people have adopted are the ones they have decided to use and are happy with is misleading. Many people only use applications by necessity and, in some cases, may decide against using them if given the possibility. Reasons include explicit constraints, social forces, and inertia.

Explicit constraints   Explicit constraints are very common in work practices. This is the case for platforms designed to structure processes (e.g., enterprise resource planning systems or workflows) but also for office applications. Typically, it is not possible to avoid using email in many jobs, and a particular email client may be explicitly or implicitly imposed—for example, via specific training sessions or a warning that the technical team will only manage issues related to this particular application. As other examples, the actual or hypothesized benefits of collaborative technologies led some employers to impose a change from email to platforms such as Slack, and many companies or institutions impose online shared agendas.

Social forces   The role of social forces is very obvious in how a normal life ("normal" being used in the sense of a life that conforms to the current norm) more or less obliges young—and not so young—people to use specific communication and social networking applications. From a technical perspective, individuals can of course decide to use the applications they want, and the technical offer is large, diverse, and generally free of charge. However, these applications create specific channels (interactants must use the same application). A basic pattern is thus that people decide—or are socially obliged—to be part of groups or social networks in the real sense (i.e., a given community of people) and, as a consequence, are obliged to use the application that these groups or networks already use. As we will see in section 5.4, this may conflict with other engagements and raise important issues.

This pattern is now so commonplace that many users are not even aware of it. However, compare this situation with email technology, which emerged in a different social and economic context: Emails can be sent or read with any email application, which allows us to use the application we prefer. In direct contrast, the technologies promoted by big tech companies such as Facebook/Meta intrinsically link the social structure (the group, the social network) and the technology. People may of course collectively discuss the pros and cons of different applications before creating the group, but this is not the most frequent pattern, and, once the group is created, newcomers have no choice and the group cannot easily be moved to another application.

Incidentally, this has little to do with technical constraints. In euphemistic terms, a business-oriented tech company has little interest in providing inter-operation and/or data exportation tools that would allow users to use other applications. Another good example is how avoiding formatting issues when exchanging docs with colleagues or relatives more or less imposes the use of Microsoft Word.

Social forces thus often lead people to use applications that they did not actively and freely select and, in some cases, that they may not have used if they had been given the choice. Most young people use the applications that their friends use, not the ones they technically prefer. And many people do not decide to use Microsoft Word or PowerPoint for reasons of personal pref-erence, but rather to avoid formatting issues and, more generally, to avoid risks by remaining mainstream.

Inertia   Finally, another reason why some people use technologies that they did not actively choose is inertia. As an illustration, Haraty et al. (2016) stud-ied the technologies that individuals use for personal task management. They found that, in many cases, an individual used one application or another (e.g., a particular email client or calendar) for the very basic reason that it was preinstalled on their computers.

The fact that skilled and/or proactive users can and generally do select the applications they use should thus not make us oblivious that many people only use the applications to which they are given straightforward access. Immediate availability is more important than the characteristics of alternative applications, which are actually not considered. This pattern is part of the rationale for the technical and legal fights related to how appli-cations are preinstalled and established as the default application to open documents or navigate on the web, and any changes made to this default by users is automatically (authoritatively and exasperatingly) overruled when the applications are updated. The way big tech companies absorb successful applications and interoperate them with their own (e.g., connecting email and drive services) also contributes to such unwilling and/or unaware uses of technologies.

Synthesis   Users do not necessarily actively *decide* to use the technologies they use and, in turn, appropriate, and appropriation is thus not necessar-ily correlated to a decision to use: One can exist without the other. We all turn technologies that have been imposed on us or are simply available into personal instruments that we fit into our practices and use constructively. Similarly, the fact that we freely decide to use a given technology may create a positive context, but this does not mean that we will appropriate it.

Moreover, as illustrated by the intrinsic social network–technology link, decisions to use should be regarded with caution. In some cases, stating that one wants to use an application may actually reflect an intention to benefit from a service that is only perceived via a known application.

The point made here is thus not that the time gap between adoption intentions and actions may lead the actual usage to differ from expectations, as nevertheless pertinently mentioned by Bagozzi (2007). The point is that the intention–behavior linkage must be studied much more carefully. Let's explore this further.

Appropriation Relates to Users' Idiosyncratic Perception of the Actual Use

The second reason for the necessity of revisiting the notion of usefulness is that, as defined in technology acceptance models, PU captures how users *foresee* the outputs of using the technology and/or make inferences from short duration uses such as a usability test, a training session, or a first contact. Appropriation, if any, takes place once the application is used *in practice*: In contrast to the situation prior to adoption, users know how they use the technology and the resulting outputs.

Although this point could lead to the conclusion that adoption is related to perceived usefulness and appropriation is related to actual usefulness, this formulation is misleading: It is not the actual gain or loss of effectiveness for the tasks at hand that plays a role, but rather the user's perception thereof. What is at play with appropriation is the *user's perception of the actual use benefits*, which is subjective and idiosyncratic, and may differ from what is "objectively" measured, by third parties, against a set of performance criteria.

We have already seen an illustration of this point with the use of email for personal information management. Fragmentation issues (Whittaker et al., 2006) and uses of multiple devices such as laptops and smartphones (Cecchinato et al., 2016) generate important difficulties. Nevertheless, studies show that some users, while aware of ready-to-use applications or features, either do not switch to these means or cease to use them (Haraty et al., 2016; Singh et al., 2013).

We can make sense of this phenomenon by taking an activity perspective: What is at stake for users is the task they address via the technology, and not the technology as such. People who are interested in technologies and actively considering them may replace one application with another because the latter is "better." However, most people are only interested in the tasks they want to achieve. When they conduct these tasks in a way that they *perceive* as satisfactory, why should they change their ways of doing things? More generally, people do not necessarily need or want to consider their own efficiency as such, and it is difficult to objectively evaluate the latter. Although conscious or unconscious resistance to change is not seen as very positive, it is a rational behavior when seen from a short-term cognitive-cost perspective. Finally, there is often a hill-climbing effect. When people develop idiosyncratic strategies to improve the efficiency of their practice, this success tends to reinforce their perception of being efficient and skilled. The users who declutter their inbox every day, place emails in files, or resend them to push them up to the top of the list do perceive these actions as improving the way they use technologies to remember things to be done, despite their strategies being far less efficient than the use of a dedicated technology. As another example, when working side by side on a word processor with another individual, we are often surprised by how he/she uses the editing or features actions, which we may consider to be indeed far less efficient than our own uses.

Another important phenomenon is that "initial patterns of using the technologies [congeal] quickly, becoming resistant to change despite ongoing operational problems in the use and performance of the technologies" (Orlikowski, 2000, p. 411). We will come back to this notion of pattern and its psychological basis in chapter 4.

What appears here is that if adoption is related to perceived usefulness in the sense of foreseen usefulness, appropriation is related to users' idiosyncratic perception of the actual use. In other words, the usefulness of the considered technology as used in practice must be considered *from the actor's perspective*.

This conclusion raises the question of the processes underlying perception and meaning-making mechanisms, which is studied in section 2.4. It also raises the question of the user's actual task—that is, the one for which he/she considers the usefulness of the technology. This is an important point because this task may differ from designers' expectations, and, in some cases, the user him/herself may not be really aware of his/her actual task. Let's study this point in more detail.

Tasks May Be Unexpected and/or Unconscious

When adoption-oriented works focus on usefulness for an *expected* task, namely the one for which the technology has been designed, appropriation is related to the usefulness of technologies for *actual* tasks. A frequent pattern is that users adopt technologies to achieve the expected task but also use it later on for others—for example, turning their email application into a personal information management system. However, a less intuitive phenomenon is that we are often engaged in simultaneous expected and unexpected (and, in some cases unconscious) low- and high-level tasks.

Use of technologies for unexpected tasks   Different labels have been used to describe how people use technologies for unexpected tasks such as creative use, repurposive appropriation (Salovaara et al., 2011), or catachreses (Rabardel, 2001, 2003). Rather than a formal categorization that would impose the definition of criteria, let's illustrate different cases.

How a device may be used for unexpected tasks is well illustrated by the use of digital cameras. Salovaara et al. (2011) list creative uses of cameras such as a mirror, a map (photo of a map), a note-taking device, a scanner, data storage, a lamp, an instruction manual (photos of step-by-step instructions), or a periscope (e.g., in a concert). Since the date of this study, some of these uses have turned into explicit smartphone features, and many others have probably emerged. For instance, Karanasios et al. (2021) note that smartphones have become a tool against police violence.

How an application may *contaminate* other tasks than those for which it was designed and, in some cases, become a means for these tasks, is well illustrated by the use of spreadsheets. Spreadsheets are designed for tasks such as calculations, financial accountancy, and statistical operations. However, Dourish (2017) shows how the anticipatory properties of spreadsheets sometimes become a means to guide discussions, and highlights that their structuration properties are a way to reuse (and thus reproduce) patterns. As an output, spreadsheets often influence or even structure activities such as presentation, coordination, decision-making, or archiving, which are all activities for which spreadsheets were not originally designed.

Finally, the use of email as a task manager device may be seen as an example of a *side effect*, although it also admittedly reflects contamination. The fact that the inbox acts as a to-do list is built on (1) the application properties (making a list of emails salient) and, most importantly, (2) use characteristics (people always keep an eye on their inbox, and many of the tasks to be remembered are mentioned in emails). However, users are also active: They implement their own management strategies to deal with these to-do lists, using filing, searching, sorting, or tagging features. Sending an email to oneself is functionally a nominal way of using the application, although it is rather ironic to send ourselves our own reminders via distant servers. Actually, the output of such behavior (having an email with a specific subject or content in the inbox) may be obtained by cheating the system: The user can simply write the email and move it from the drafts folder to the inbox without using the features that serve the basic purpose of the application—that is, sending or receiving emails.

As mentioned in chapter 1, such function creeps remain the exception rather than the rule. The most common answer to "What do you use your email for?" is: "Communicating with others by sending and receiving emails." Nevertheless, in addition to being highly illustrative of the appropriation phenomena, such innovative uses clearly show that the usefulness of the technology must be addressed by considering the user as the analytical entry point, rather than the technology design rationale, and his/her actual activities.

As a way to further illustrate how focusing on user activities allows us to make sense of (sometimes odd) uses, let's take a more general perspective to our remembering-things-to-be-done running example. Studying how creativity helps negotiate everyday life, Wakkary and Maestri (2007) conducted an ethnographic study of three families. In their words, "resourcefulness, the creative re-use of artifacts and the physical surroundings, is a building block for everyday design" (p. 164). One of the occurrences they identified was the solution found by one of the studied persons while facing the unexpected need to find a place for a bouquet of flowers. The person used another recipient to temporarily free a used vase, then placed the recipient on the fridge as a reminder to deal with the situation at a later time. The recipient on the fridge plays the same role as the email in the inbox: It is a visual reminder. Dedicated empirical analyses of how we remember things confirm that placing objects in locations where one is sure to encounter them in daily routines (e.g., in the kitchen, at the front door or beside the car keys) is a very usual general strategy (Brewer et al., 2017).

What appears here is highly illustrative of the account of appropriation proposed in this book. We have some more or less general needs (here, remembering things). We respond to these needs by developing activities and strategies that, while contextual, may be influenced and/or shaped by previous reactions to similar needs (in other words, we reuse previously shaped psychological constructions). Keeping an email as a to-do and putting something on the fridge implement the same abstract strategy: Knowing that an odd situation will make us wonder what is happening, we artificially create odd situations to remind ourselves to think of something specific. The most classic example is the knot in the handkerchief (at least, for the generation who used the latter and/or for Vygostky readers).

When considering user perception of the usefulness of an application, what is to be considered is thus the actual user tasks. For instance, in many situations, a collaborative platform such as Slack is arguably much more efficient than email exchanges; however, for some users, the fact that their email is (or is perceived as) an efficient reminder may hinder any evolution toward forum- or channel-based platforms. Making sense of such cases is thus core.

Engagement in simultaneous domain (low-level) and personal (high-level) tasks Another reason for our use of technologies for unexpected tasks is that we are not actually aware of all the tasks we engage in. Although it is a difficult phenomenon to identify and, sometimes, to acknowledge, we also address unconscious tasks.

Analyzing the rationale for how some people manage their emails is very illustrative. For instance, email overload is an issue for many workers, which leads them to explicitly address an overload management task (Dabbish & Kraut, 2006; Grevet et al., 2014). This task may be addressed by defining email filters, which allow actions such as automatically moving emails to a specific folder, tagging them, or deleting them (see chapter 6). However, Barley et al. (2011) show that the fear of missing an important email prevents many users from using filters and other features designed to screen or reduce email volume, which in turn accentuates the overload issue. Hanrahan et al. (2016) added to this observation the understanding that what is actually at play for some users here is the defense of their reputation for efficiency, which is related to how promptly they deal with important messages.

What this example illustrates is that we have interwoven low-level (domain) and—possibly unconscious—high-level motivations (e.g., defending one's reputation), and what we use the technology for may thus have a multi-level response. When considering user perception of the usefulness of an application, the possible interplay of low-level and (possibly unconscious) high-level tasks must thus be taken into account. As another example, social networking apps are used to send/receive/share different materials and also to carry out higher-level tasks such as expressing aspects of the user's identity or gaining social capital, although the users may not be aware of this.

The fact that such general and possibly unconscious forces play a role has been empirically studied by works conducted in the TAM research tradition. Examples of identified forces are: anxious or emotional relations toward a task or a technology, affects or personal expectations as related to self-esteem or sense of accomplishment, or compatibility with existing values (Venkatesh et al., 2003); result demonstrability—that is, if and how results of adopting the technology is observable and communicable to others (Karahanna et al., 1999); and perceived transition costs or unconscious factors such as the use habits of the incumbent system, which may negatively impact the perception of the ease of use and relative advantage of new technologies (Polites & Karahanna, 2012).

In a nutshell: What is at play with appropriation is the user's perception of the actual use benefits for the tasks that the user considers. These tasks may or may not correspond to designers' expectations, and the user's perception may be idiosyncratic. Hypothesizing that users perceive the usefulness of an application in respect to the initial design objectives of the application is thus too simplistic and, in some cases, may prove to be misleading.

2.2   The Role of User Ecologies of Artifacts

In the previous section we considered the question of the usefulness of a technology. Although I argued that this must be studied by considering the user and his/her tasks as the analytical entry point, considering the usefulness of a technology is nevertheless a techno-centered perspective: The focus is on the technology.

Let's now reverse the perspective—that is, consider how people address a particular goal and, in this context, use technologies. Taking such an activity-centered perspective immediately highlights the fact that people may (and indeed do) use several artifacts to achieve their tasks.

The implication is straightforward: How users perceive the usefulness of a technology and use the latter cannot be understood by focusing solely on this technology, and a more holistic perspective is needed. This may be addressed via the notions of ecology of artifacts and places.

Considering Users' Ecology of Artifacts

The notion of *ecology of artifacts* was introduced to describe the set of interactive artifacts that a person owns and/or has access to and uses (Jung et al., 2008). The reference to the notion of ecology and its biological roots reflects the idea that the technologies we use form complex and dynamic networks. The authors' empirical study and many others reveal how the roles that users attribute to each technology are influenced by the roles they attribute to other technologies. In particular, users relate technologies to each other because they share common features or exchange data, but also due to their perceived idiosyncratic similarities or relationships.

As an example, the studies analyzing how people remember their personal tasks (rather than how they use a given technology) reveal how people generally use several items from a large and diverse set of artifacts such as paper planners, pieces of paper, emails, e-calendars, alarms, word documents, Wikis, paper notebooks, e-notebooks, specific to-do-list applications, and/or Post-it notes (Haraty et al., 2016). Attempting to understand how and why people appropriate one technology by considering it in isolation from all others is thus misleading: The use of every specific technology is shaped, or a minima impacted, by how the other means are used.

As another example, most young people address the high-level task of socializing via multiple means and attribute different purposes to social networking applications. For instance, the study conducted by Boczkowski et al. (2018) reveals that, for the users in question (young Argentinians) and as observed at the time of the study, WhatsApp is used for sharing information quickly with friends and family, Facebook for disseminating content more widely, Instagram for "careful and stylized constructed visual portraits of everyday life," Twitter to get news and comments, and Snapchat for instantaneous and fun communication with friends (p. 255). This implies that the use of each platform is related to if and how the others are used (and, of course, social networking apps are but one of the means that young people use to socialize, along with other digital and nondigital strategies such as dress codes).

Ecologies of artifacts are dynamic (Bødker & Klokmose, 2012): Some technologies may be or become frequently/infrequently used, new ones may become part of the ecology (or not) for a variety of reasons including functional and nonfunctional aspects, and others may cease to be used and be removed from the ecology.

The role of nonfunctional aspects is illustrated well by the way people attribute different purposes to social networking applications. With respect to the specific usages that emerge and how they evolve, Boczkowski et al. (2018) pertinently highlight that "neither the uses nor the meaning constellations are an expression of each platform's affordances: There is nothing that technologically discourages silly pictures on Instagram, serious ones on Snapchat, or exclusive content on WhatsApp—to the contrary, users might enact each of these options on all respective platforms" (p. 246). The uses that crystallize are mainly shaped by the socially constructed understanding of what is acceptable and desirable communication on the different platforms.

Although the notion of ecology of artifacts is often used to refer to devices, the fact that features of different devices may overlap and/or interoperate means that we must consider the features level and not just the devices or applications levels. For instance, as we saw, managing emails on both smartphones and laptops raises specific issues (see section 1.2). This implies that the way people use email applications is impacted by the devices they use (e.g., laptops, smartphones, or smartwatches), by the specific features of the applications they use on these devices (e.g., simply accessing or downloading emails, or the filtering and tagging tools), and by the way they deal with the possibilities and the issues that this complexity entails. Similarly, individuals using several social networks and/or game platforms communicate with the same people via these different means.

Here again, work practices or social structures also play a key role. Continuing with the email example, studies reveal how email management is influenced by broad aspects of work such as the level of responsibilities, dependence on other individuals' activities, or work-home boundaries (Haraty et al., 2016; Cecchinato et al., 2015; Dabbish & Kraut, 2006).

From a theoretical perspective, the dynamics of users' technical ecosystems may be addressed in the light of the actor-network theory (ANT) approach, whereby the adoption or rejection of technologies is addressed as a modification of the networks of relationships relating actors and artifacts (Latour, 2005). However, I will not be taking this path. The rationale is that ANT addresses the human and nonhuman (technological) nodes of such networks in a symmetrical way. In direct contrast, the perspective I develop in this book stresses that humans are intentional actors and, as such, are guided by needs and motives.

Considering Places Rather Than Spaces

The second implication of the fact that users simultaneously use and articulate several different resources is that the organization of these resources may play a specific role. This issue can be considered in the light of the place/space dissociation.

The distinction between *space* and *place* was introduced to better understand collaborative work settings (Harrison & Dourish, 1996; Dourish, 2006). Space is the structure of the world that shapes, constrains, and enables certain forms of movement and interaction, and "a place is a space which is invested with understandings of behavioural appropriateness, cultural expectations, and so forth. We are located in 'space', but we act in 'place'"; "Space is the opportunity; place is the understood reality" (Harrison & Dourish, 1996, p. 69).

Nouwens et al.'s (2017) analysis of the use of different communication apps nicely illustrates the importance of considering the places that emerge. Their empirical study reveals that how users decide to use one app or another, and how they use it, stems from how they create idiosyncratic communication places that each have their own unique membership rules, perceived purposes, and emotional connotations. Typically, family, friends, or colleagues are usually located in different places. The technical features and specificities of apps (for example, listing contacts or signaling a user's online presence) either support or conflict with these places' properties or, in other words, with the idiosyncratic identities that users have given to these apps, which has an impact on how they are used. For instance, these authors found that the emotional connotations attributed to some apps led users to manage issues such as not wanting to see a contact at the top of the contact list or skipping from one app to another as app identities and/or contact statuses evolve.

As another example, the study by Retore and Almeida (2019) reveals issues with the "last seen" feature: Keeping this awareness information may be a way to "[let] my mom know that I'm alive" but may also cause situations such as "I feel like I'm being watched" (p. 259). In other words, the psychological need to maintain independent social spheres is in potential conflict with the technology features (Binder et al., 2009), leading to unexpected impacts on uses.

In a nutshell: As appropriation stems from user activities, and activities are generally mediated by several digital or nondigital artifacts, making sense of why and how people appropriate technologies requires the consideration of (1) their ecology of artifacts and its dynamics and (2) the idiosyncratic functional organization that users may have developed.

2.3 The Role of Mediation Mechanisms

The two preceding sections clarified that what must be considered is the user's idiosyncratic perception of his/her actual uses as a means to achieve his/her activities, and that the analysis must embrace the individual's full set of means. Let's now study in closer detail what *using the technology to conduct one's activities* means, particularly when taking a psychological lens, and what the implications thereof may be.

Mediation and Activity Theory

It has been an important breakthrough to understand that, rather than considering computers as something that people work on, they should be considered as something people act through (Bødker, 1991). In other words, computers or smartphones are some of the *mediators* in our interactions with the world.

Taking this perspective leads us to address human activity as a purposeful, mediated, and transformative interaction between human beings and the world. The purposefulness of activity stems from intentionality: What is core is that people have needs that they try to meet. As suggested by authors such as Bødker and Klokmose (2011), activity-centered analyses of the use of technologies should thus be structured by considering what they call the "why," "what," and "how" considerations: the motivational orientation (why?), the goal orientation (what?), and the operational level (how?). These authors later extended this analytical means to embrace user ecology of artifacts (Bødker & Klokmose, 2012).

Actually, I have adopted this general motivation-based perspective right from the first pages of this book. For instance, the email example may be reframed in this way: Why?—because people need to remember things to be done; What?—they manage to make emails highlight the existence of a task; How?—they manage their inbox in a way that draws their attention to the emails that indicate tasks to be done, which includes creating new ad hoc emails.

Analyzing how technologies act as mediators requires a psychological understanding of the phenomena at play. The specific interest of the one proposed by activity theory (AT) first and second generations is to natively consider (1) the low-level psychological mechanisms governing how and why we act and (2) their high-level social and cultural dimensions. As we saw, both play a key role in appropriation.

AT is a descriptive conceptual framework arguing that human development and psychological functioning are rooted in social processes and must be addressed in light of the historical development of the individual in society (Leont'ev, 1978). At the very heart of AT is the idea that relationships between humans and the world are not direct: They are mediated by socially and culturally constructed objects. Mediation as addressed by AT may be viewed as "a complex system of objects and structures, both material and immaterial which serve as mediating means embedded in the interaction between human beings and the world and shaping the interaction" (Kaptelinin, 2014b).

In the rest of this section I will focus on how some of the AT notions can be used as a psychological substratum for making sense of part of the appropriation phenomena. Readers interested by the theory as such can find an introduction to AT, its main notions, and their interest for the human-computer interaction (HCI) field in the online presentation proposed in Kaptelinin (2014b), the reference books Nardi (1996) and Kaptelinin and Nardi (2006), or the survey proposed by Clemmensen et al. (2016). For an analysis of the evolution of AT and its first, second, and third generations, see for example Spinuzzi (2020). While AT takes its roots in the Russian sociocultural psychology tradition (~1920–1930), it later developed internationally, in Scandinavia in particular. It has been used in areas such as the study of work practices and organizations, education, or information systems.

Functional Organs

The way AT addresses mediation and the notions it proposes to do so are of specific interest for the understanding of appropriation.

First, AT posits that "the distribution of activities between mind and artifacts is always functional. It takes place only within subsystems that have specific functions, more or less clearly defined" (Kaptelinin & Nardi, 2006, p. 65). This point provides a theoretical background for the role of perceived usefulness as defined in section 2.1: Understanding the uses of a technology requires focusing on the functional drivers thereof—that is, the user's actual tasks, needs, and motives, whether explicit or unconscious.

Second, the AT perspective questions the dissociation between humans and technology. Considering humans on one side and technology on the other may not be the best way to understand what happens. This resonates well with the idea of making technology our own and what it means.

Third, Leont'ev (1981) introduced a notion that captures these two points, namely that of *functional organs*. Kaptelinin (1996) proposes the following definition: Functional organs are "functionally integrated, goal-oriented configurations of internal and external resources" (p. 25). The notion of functional organs embodies the idea that technologies (or, using AT words, artifacts) are amplifiers of human capacities. In Kaptelinin and Nardi's words (2006), "Functional organs combine natural human capabilities with artifacts to allow the individual to attain goals that could not be attained otherwise" (p. 64). The classic examples include how glasses amplify vision or a cane amplifies an individual's capacity to walk. Similarly, the combination of human capabilities and email allows an individual to communicate with others and to remember things to be done, while the combination of human capabilities and Facebook or Instagram allows people to gain social capital.

The definition of functional organs proposed by Kaptelinin ("functionally integrated, goal-oriented configurations of internal and external resources") highlights three core ideas. The first is that the driver of the psychological development of functional organs is the user's goal. Second, what is at play includes both external resources (the artifacts) and internal resources (the psychological constructions that the user develops to conduct his/her activities with these artifacts). As we will study in detail in chapter 4, this includes conceptualizations (for instance, using email requires users to develop and differentiate a set of notions such as "message," "sender," or "CC") and mental schemes (ways of doing, ways of addressing situations). Lastly, these internal and external resources must be addressed together and in relation to their rationale—that is, the functional driver.

AT thus (1) stresses the importance of considering how people develop the internal resources that form the psychological substratum of their appropriation of technologies and (2) provides us with a possible criterion for considering if a technology has been appropriated or not, which is: Has the technology become a functional organ or a part thereof? This perspective raises two important questions. The first is the issue of defining if something has become a functional organ. The second question is why some artifacts become amplifiers of human capacities when some others do not. We will come back to this in the following chapters.

As a side comment, considering the "user + technology" couple as an entity raises the question of which boundary should be considered: The one separating the user and the technology that is used or the one separating the couple (user + technology) and the outside world? (Kaptelinin, 1996). The historical example of the boundary issue is the blind man and the stick; a more recent one may be young people and their smartphones.

Perceived Ease of Use and Usage Transparency

When considering appropriation, perceived ease of use must be reconsidered in a similar way to perceived usefulness—that is, focusing on the actual and in-context (rather than anticipated and in-lab) perception. The central question is how a given individual perceives and uses a given application rather than the usability of the technology as such.

This issue can be conceptualized by considering the notion of *usage transparency*, which is commonly used in HCI. Usage transparency refers to a state where the user thinks about the task he/she is doing rather than the application he/she is using. Typically, an email client is used transparently if, when one sends a message, one focuses on what we want to say to our correspondent and not on the editing or sending features of the application. In other words, one acts through the application, but this mediation is transparent. The notion of transparency has also been used in works that study Csikszentmihalyi's notion of flow in the context of digital technologies, such as Finneran & Zhang (2003) or Pilke (2004). A flow depicts the state where one is fully absorbed by one's activity with a sense of personal control, which leads to a loss of self-consciousness or an altered experience of time. The classic examples include painters or climbers. As mentioned by these authors, flow experiences with digital technologies may be generated by the use of the technology (the user is absorbed by and enjoys the use of the system per se) or by the task one addresses through the technology. An interface that is not transparent would typically prevent or break a flow experience related to the task.

Transparency is an evolving qualitative aspect of the relationship between users and systems and is related to attention rather than perception

(Kaptelinin & Nardi, 2006). At any moment, and particularly when facing an issue, we can draw our attention back to the system.

Activity theory makes it possible to understand usage transparency as follows. Human activities are oriented to motives. These activities are implemented through chains of actions, which are controlled by conscious goals. These actions are carried out through operations, which stem from the automatization of actions (Kaptelinin, 1996) and are performed without conscious thought. In the light of this theoretical framework, Bardram and Bertelsen (1995) pertinently define transparent interaction as handling the computer through operations.

In a nutshell: AT and, particularly, the notion of functional organs as an interplay of external (artifacts) and internal (psychological) resources in relation to a functional driver (user goals), provides a conceptual basis to comprehend the psychological mediation phenomena underlying appropriation, including usage transparency.

Is use transparency—that is, the fact that using the system does not generate an interfering conscious process—a criterion for considering that a technology has been appropriated? I will argue later in this book that use transparency is a likely symptom of appropriation, albeit a contingent aspect thereof. Rather, appropriation usually—though not necessarily—leads to frequent uses, with an effect on use transparency.

2.4 The Role of Perception and Meaning-Making Mechanisms

Now that we have a conceptual basis to address how technologies mediate our actions, we can move on to clarifying how we (humans) make the connection between our motives and these means. How we perceive the way we can (or should) use a technology is a multilevel (i.e., involving low-level psychological mechanisms and high-level social forces) and multi-process (e.g., perception and meaning-making) phenomenon, and it must be addressed as such. This section presents the contributions of different research traditions. First, the implications of how AT suggests to address the notion of affordance. Second, the implications of how the structuration and genre perspectives suggest to address the social aspects of perception and meaning-making and, in particular, the existence of socially conveyed ways of doing.

Inputs from the Works on the Notion of Affordance

Technically, a digital technology is an artifact proposing a set of features. These features can be precisely and unambiguously described. However, as we have seen, what is important is our perception of these features and their usefulness for us. This may be studied in the light of the notion of *affordance*.

The notion of affordance was originally proposed to address how animals live in their environment: "The affordances of the environment are what it offers the animal, what it provides or furnishes, either for good or ill" (Gibson, 1979, p. 127).

Since its introduction to the HCI field in the late 1980s, the notion of affordance is a subject of intense debate. Broadly speaking, it is used to refer to the action possibilities offered by technologies. However, major disagreements have developed, particularly with respect to the relation between affordances and their perception: Do affordances exist as such—that is, independently of whether they are perceived by users? Should the notion of affordance-as-action-possibilities be dissociated and disconnected from the notion of information-about-the-affordance? Is acquiring the ability to perceive affordances an oxymoron? Must one consider that the social context shapes or influences the underlying perception and/or meaning-making mechanisms— that is, that affordances are to some extent socially and culturally constructed? The literature on the subject is wide and controversial. See McGrenere & Ho (2000) for a compared analysis of different definitions, and Kaptelinin (2014a) for a synthesis.

Rather than make yet another tentative contribution to this general debate, I will focus on two points that are of importance to the motives-means linkage: first, the difference between *use intuitiveness* and *semantics intuitiveness*, and second, the role of activity in perception.

Intuitiveness: Use and effect   Intuitiveness—that is, the property of being directly apprehended—is generally considered an important design goal and a criterion to evaluate user interfaces. It may be addressed in the light of the notion of affordance, which was originally introduced to HCI as "the perceived or actual properties of the thing, primarily those fundamental properties that determine just how the thing could possibly be used" (Norman, 1988, p. 9). Although Norman would later explicitly reconsider this definition, it has remained a common way to refer to affordances, probably because it clearly resonates with design considerations: Digital technologies are designed for expected uses, and "good" interfaces should provide clues suggesting these uses.

However, interface intuitiveness may relate to two different aspects. One is how the artifact may be used, and the other is what effect it has. It is important to dissociate these two aspects, which differ in nature. A classic illustration of this difference is how the Macintosh interface allowed users to drag not only files but also the disk icon into the bin. Although the use of this feature was very intuitive, the same was not true for its effect: Dragging the disk icon into the bin ejected the disk but was incorrectly understood by some users as deleting the disk.

In the light of this dissociation, the fact that perceived usefulness is the core driver of appropriation means that, when considering the role of affordances in appropriation, the aspect that must be considered is the perception of what the technology enables the user to do.

This point may be further framed in terms of *handling* and *effecter affordances*. These notions were proposed by Kaptelinin and Nardi (2012) in the context of their call for regrounding the affordance notion in a mediated action perspective. The authors suggest considering that the action possibilities provided by a technology include the possibilities for interacting with the technology (handling affordances) and the possibilities for employing the technology to make an effect on an object (effecter affordances). In the authors' words: "Together, [handling and effecter affordances] define instrumental technology affordances as possibilities for acting through the technology in question on a certain object" (p. 972). Typically, when attention is driven to the possibility of tagging important emails or using emojis to communicate one's feelings via a communication app, what is at play here is an effecter affordance.

This clarification opens the question of what mechanisms are at play in the perception of action possibilities as means. Let's explore this very important point further.

Perception of action possibilities as means Taking the technologies-as-media perspective introduced in section 2.3, technical features are not affordances as such. As put forward by Bærentsen and Trettvik (2002) and many other authors, they *become* affordances if and when the user relates them to his or her activity. In other words, affordances emerge within the interactions that users develop with the technology while achieving their tasks.

Let's first consider a classic (nondigital) example. I want to open a bottle of wine and do not have a corkscrew. An instrumental way to deal with the issue is to open the kitchen drawer or the toolbox and look for something that may help. In other words, I look for something that has a removing-the-cork-obstacle affordance and look for this in a place that has a contain-objects-that-may-have-a-removing-the-cork-obstacle-affordance affordance (nested affordances [Gaver, 1991]).

What is happening here is that I am seeking a solution to my problem, and my problem, or, rather, the way I usually address this problem, is what drives my attention when I look through the contents of the drawer. My goal defines what is important for me, and I am waiting for my perception mechanism to detect an object that is suitable in this context. This resonates with the fact that in Gibson's perspective "The purpose of perception is to efficiently obtain meaningful information, that is, information that has significance to acting in the environment" (Kaptelinin, 2014a). For a more in-depth discussion of Gibson's perception perspective and Hartson's (2003) notion of functional affordance (the usefulness of a system function) and its discussion based on the corkscrew example, see Kaptelinin (2014a). As a side remark, if I find something in my drawer that solves my problem, this may be referred to as a one-off appropriation (see chapter 3).

The way the purpose of action orients and shapes perception has also been highlighted in the context of AT. For instance, Ochanine's (1977) empirical analyses led him to dissociate the *cognitive image*, which corresponds to all the accessible properties of an object, from the *operative image*, which refers to what is useful for a given goal. This dissociation reveals the role of the operative image in the process of picking information and computing in and for action. The central importance is not the object substance but its function for the user.

An important point made by these works is that what the user will perceive as an option is not solely driven by the characteristics of the artifact that is usually used; the associated way of doing is also part of the picture. The large screw I discover in my toolbox is likely to attract my attention: It looks like a corkscrew and may be used in the same way. An extra-narrow clamp does not look like a corkscrew but may catch my eye if I am looking for a means to pull out the cork—that is, the way of addressing the task that underlies the use of corkscrews. And if my conceptualization of the removing-the-cork task is limited to the extraction technique, I may not perceive the tubular item that could be used to push the cork inside the bottle. Different factors may also play a role here, including sociocultural aspects. Typically, for a Frenchman, the cultural factors associated with the opening of a bottle of wine may occlude any perception of means that are efficient but would shock fellow diners.

Let's now reframe the remembering-things example in this light. This need/goal/task leads people to (more or less consciously) search for mediating artifacts. This search is what may make users perceive as affordances (1) the different means that designers have created for this purpose (e.g., paper Post-its, electronic agendas, or task managers) and/or (2) features or artifacts that were not designed for such a purpose but, for some reasons, are idiosyncratically perceived as having a reminding effect. This is the case for the "keeping emails as to-dos" practice identified by the empirical analyses of email uses. As an output, the inbox becomes a functional organ or, rather, part of our "remembering" functional organ, together with our biological memory and other elements of our ecology of artifacts such as online or paper-based calendars.

Good designs make us perceive relevant effecter affordances and direct our perception mechanisms toward efficient means. For instance, at the time of writing this book, for many years now most if not all laptop users, when confronted with a new application interface, have perceived the application "menu" as an affordance for discovering what features could help them address their tasks (nested affordances). In other words, they act (here, browse the menu options) to give their perception mechanisms a chance to identify something (here, an option) that matches their needs. Although this is basic design practice for websites and apps, many users however still do not perceive the "burger icon" (≡) as offering a similar affordance.

If and how our perception mechanisms detect a feature as a possible means is related to our conceptualizations—that is, the notions we use to make sense of the world and of our activities. The burger icon is a notion for some users and not for others. When they browse the interfaces of email clients for the first time, most people detect features such as "send email" or "attach document" because they are familiar with the underlying notions and can infer what is at play. This may not be the case for options such as "set encoding to UTF-8." Similarly, many of the features proposed by complex applications such as Photoshop or GIMP are not affordances for anyone who is not a photo specialist. Features based on notions such as "layer," "gradient," or "sharpness" cannot be detected as means if one has no idea of what these notions mean.

I mentioned in chapter 1 that as smartphones are increasingly used for almost every need, the first move of users' facing a digital task is often to browse their apps and/or favorite apps repositories in the hope of finding a solution. In a certain way, the smartphone is appropriated as a toolbox—that is, as something that has a contain-potential-solutions-to-a-wide-set-of-digital-problems affordance (here again, nested affordances).

Implications   With respect to understanding appropriation, the way these analyses make sense of affordances has a core implication: Our perception and interpretation of the technology features we consider and, thus, their perceived usefulness for us, are shaped by our actual needs, our usual ways of doing, and their underlying conceptualizations. Considering these elements is thus core.

This point once again supports the perspective arguing that the fact that users fail to perceive some of the means offered by designers, or perceive them in an expected way, is not simply a matter of "good" and "bad" designs (although there are of course good and bad designs) and/or "skilled" and "unskilled" users. The perception and interpretation of technology by users are driven by their objectives and usual ways of doing and, thus, are highly idiosyncratic. When usability studies address nominal settings (the achievement of the tasks for which the technology has been designed, the expected conditions of uses, and so on) and users in the plural (i.e., in general), appropriation stems from the actual setting (the actual task) and the individual (e.g., his/her conceptualization of the activity). As mentioned in chapter 1, user perceptions and subsequent uses may thus be consistent or inconsistent with the design "spirit"; they may or may not be efficient, but they are not good or bad. The very idea of "correct interpretation" is debatable.

Considering that users' perception and interpretation of the technologies they use stem from user needs, tasks, usual ways of doing and underlying conceptualizations (1) clearly differs from the technology-based perspective in which the user is considered to first and foremost perceive and interpret technologies on the basis of their characteristics but, however, (2) does not mean that technologies are "neutral" or that their perception and interpretation is a purely individual process. First, the technical characteristics of technologies do of course play a role, if only because different technologies present different features. Moreover, the design of these features and, in particular, what they leave open to users, is of great importance. I will come back to this in chapter 6. Second, as highlighted by activity theory and other research traditions (see the next sections), the mediators elaborated within a sociocultural context and, in particular, the technical artifacts, convey socially elaborated ways of doing things via both their design and the associated knowledge and discourses. For instance, the corkscrew crystallizes the socially accumulated experience of how to address the "opening bottles" task, and the email application crystallizes the socially accumulated experience of how to exchange asynchronously with distant correspondents. When we encounter these technologies, we thus also encounter the socially elaborated ways of doing that they originate from and convey, and recognize them.

The role of socially shaped ways of doing is therefore central to how we perceive, interpret, and thus appropriate technologies. These ways of doing (1) structure most of the activities within which we use the technologies (e.g., how we exchange with colleagues or friends, write documents, or conduct our professional tasks) and (2) are more or less inscribed within these technologies: The proposed features and how they are presented reflect these social constructions.

In the next sections, I will present two research traditions emphasizing this dimension: the structuration theory and the genre approach. In chapter 4—that is, in the context of the study of psychological (developmental) mechanisms—I will present another approach of this topic, namely how the instrumental genesis theory addresses crystallized ways of doing by considering instruments as mixed entities combining technical and psychological dimensions.

Inputs from the Structuration Theory Perspective

The structuration theory (Giddens, 1984) proposes a general understanding of how social systems are created and reproduced through social structures and human agents. The notion of structure, as used here, refers to the often tacit and/or less easily observable rules of social systems (e.g., status hierarchies, organizational knowledge, or operating procedures) and the resources thereof—that is, the observable patterns of relationships among individuals and collectives. This theory has been considered as a key element of the understanding of organizational communication (and has also received some criticism).

Several important works addressing technology appropriation have built on this theory such as the adaptive structuration theory (Poole & de Sanctis, 1989; de Sanctis & Poole, 1994) or the analyses conducted by Orlikowski (1992, 2000) and Vyas et al. (2017). This is not the case of this book. As explained in chapter 1, I consider that fully understanding appropriation and developing an urgently needed emancipatory perspective requires an activity-centered and developmental analysis. Nevertheless, the aforementioned works have made important contributions, which I will summarize here.

Structures impact perception and convey ways of doing   It is clearly obvious that structures lead to established and sometimes institutionalized ways of conducting some tasks. Structures, which are the result of previous human actions, both enable and constrain actual actions.

The structuration theory, however, also puts forward the interplay between structures and agents. The actors are knowledgeable and reflexive. The way they interpret structures is influenced by their own characteristics (e.g., experience or motivation) and the characteristics of the context (e.g., social relations or local resources). In other words, there is a recursive relationship between the activities and the social structures that are the medium and outcome of these activities. As pointed out by Orlikowski (1992), designers embed in the technology rules and norms related to the work being automated. Meanwhile, users appropriate the design by assigning shared meanings to it, which plays a role in how the technology characteristics and embedded rules/norms influence their task execution. Both designers and users are influenced by their institutional conditions of interaction with technology (e.g., practices, norms, or resources).

With respect to our topics of interest, this understanding of the role of social structures sheds some light on how technologies embed or reflect beliefs about the considered activities and the role of technology to address them (Orlikowski, 1992). The associated crystallized ways of doing and using stem from the interaction between technology and organizations, and not just from the technology alone. They are conveyed via the technologies' structural features but, also, via their spirit—that is, the "official line" regarding how to act with them (Orlikowski, 1992).

As obvious examples, enterprise resource planning systems (ERPs), workflows management systems, or task management applications convey more or less explicit "best practices" (as identified by the designers). They come with expectations of how their users will use their features but, more generally, will address the task, including for the unspecified aspects thereof. This is also the case of how collaborative platforms suggest to use online shared calendars or distribute tasks structured via "to do," "doing," and "done" labels.

As a very different example, Sun et al. (2020) analyzed in the light of the structuration theory the issues faced by massive open online courses (MOOCs), which suffer from extensive dropping out. The structuration analytical lens highlights that a core aspect of the MOOC spirit is collaboration, and the technology reflects the belief that learners will engage in joint efforts, rely on each other, and develop social relations. The benefits of MOOCs are thus highly dependent on if and how the participants develop a shared and faithful appropriation of the technology, which is only the case for some of them. Such cases may also be addressed by considering what de Sanctis et al. (2008) call the consensus of appropriation—that is, "the degree of agreement among group members regarding how [the technology] should be used and how it fits the group's work" (p. 558).

The structuration theory has also been used to make sense of technologies' affordances (Vyas et al., 2017), and, while the approach is significantly different from the activity-centered one I emphasized in the preceding section, the implications come close. Coherently with the structuration theory, Vyas et al. suggest addressing affordances as a dynamic emerging relationship between people and their environment that is impacted and possibly shaped by social and cultural factors. In the authors' words, "Designers may have inscribed certain meanings and uses of a technology, but it is the way users use that technology in a certain cultural and social situation determines its affordance" (p. 118). "From the structuration theory perspective, a specific format of technology use (technology-in-practice) determines what the technology affords" (p. 121). Building on these premises, the authors propose to address affordances by considering three levels (single user, work group, societal), and they identify four main factors that may affect the emergence of affordances: (1) the technological conditions, such as the proposed features, (2) the cultural conditions, for example, beliefs and values, (3) the power conditions, such as relations of dominance, and (4) the interpretive conditions—for example, attitude toward the technology and the meanings attached to it. The proposed illustration is the commonplace appropriation of fax machines as photocopiers. Although it does not change anything in terms of the technological conditions, cultural conditions such as organizational norms or power conditions (e.g., proximity with somebody in a higher position) may impact this appropriation process.

As a side remark, this means that the appropriation of a technology is, in some sense, the appropriation of the structures of rules and resources that it conveys (Poole & de Sanctis, 1989; de Sanctis & Poole, 1994). In other words, technologies convey social structures, and the uses that develop lead to an evolution or reinforcement of the institutional structures. When taking an emancipatory perspective, this gives a specific connotation to what "making a technology our own" may mean.

Specificities of individuals' uses: Technology in practice versus artifacts

While the structuration perspective puts emphasis on the role of structures and their reproduction, Orlikowski (2000) stresses that it is when the properties of the technologies introduced by the designers are "mobilized in use" that they structure human action. In other words, structures emerge in action.

This perspective leads this author to make a point that I think is core: "Rather than emphasizing technology and how actors appropriate its embodied structures, [start] with human action and [examine] how it enacts emergent structures through interaction with the technology at hand" (Orlikowski, 2000, p. 407). From a conceptual perspective, the logical implication is to dissociate the technological artifact and the *technology-in-practice*—that is, the enacted structures of technology use or, in her words, "the sets of rules and resources that are (re)constituted in people's recurrent engagement with the technologies at hand" (p. 407).

An illustration of technology-in-practice and the interplay of appropriations by individuals and groups can be found in the study of Slack appropriation reported by Retore and Almeida (2019). As part of the adaptation of the application to the work routine, one of the users reacted to the high number of documentation requests by configuring an automatic response with the relevant files attached. This end-user tailoring both improves efficiency (automation of repetitive tasks) and changes the organization of work. A new task appears, namely that of keeping the attached documents in the automatic replies up to date. As pointed out by the authors, the very meaning of the application evolves, moving from a space for information exchange to a reference to find files and data. This meaning stems from the context-specific social construction of the work practice.

As another example, Schmitz et al. (2016) argue that the structuration perspective allows to make sense of how, on the individual level, some workers develop new uses of technologies (e.g., smartphones) and create new dynamics. They propose to dissociate "exploitive technology adaptation" (user extension of the technology capabilities consistent with its spirit, which may or may not include technical customizations), "exploitive task adaptation" (user modification of existing task processes consistent with their current structure and objectives), "exploratory task adaptation" (user transformation of task processes generating new objectives), and "exploratory technology adaptation" (user extension of the technology capabilities that may not respect the original technology spirit). The aforementioned Slack example could be analyzed in terms of exploitive/explorative technology and task adaptation.

Let's now study the genre perspective, which also emphasizes the role of social constructions. While sharing some points with the structuration perspective, this research tradition sheds specifically useful light on appropriation processes.

Inputs from the Genre Perspective

As writing is a powerful and widely used mediational means for organizations and institutions, understanding "writing in use" is crucial (Russell, 2009). The genre research tradition, which mainly developed from the US tradition of rhetorical analysis and Bakhtin's notion of genre (see Bazerman, 2013 for details), addresses this issue in detail.

Genres have been defined as typified rhetorical actions based in recurrent situations that have been routinized over time (Miller, 1984, 2015). Basic examples of written genres that may be found in work practices include proposals, annotations, memos, reports, or forms.

Taking a genre perspective on the use of digital technologies allows the identification of such recurrent patterns. Examples include email genres such as release notices or team reports (Yates et al., 1999); PowerPoint genres in terms of purpose, content, form, participants, time, and place (Yates & Orlikowski, 2007); or, on the web, "homepage," "news discussion," or "product reviews" genres (Askehave & Nielsen, 2006; Shi et al., 2020).

Although initially focusing on texts, the genre approach thus provides general insights into how artifacts combine with socially constructed ways of doing things. This is what I will focus on.

Genres as traditions of addressing tasks/using artifacts The way Bazerman (1994) introduces the genre research tradition resonates well with the more general question of how we perceive mediational means and how we can or should use them that I developed above. In his words, "Genre theory . . . has been concerned with the development of single types of texts through repeated use in situations perceived as similar. . . . Individuals perceive homologies in circumstances that encourage them to see these as occasions for similar kinds of utterances. . . . Genres, in-so-far as they identify a repertoire of actions that may be taken in a set of circumstances, identify the possible intentions one may have . . . [and] embody the range of social intentions toward which one may orient one's energies" (p. 69).

What is specifically interesting for the understanding of appropriation is how the genre tradition has demonstrated the existence and role of conventionalized ways of solving particular tasks. Considering writing as one case, Russell (2009) defines genre as "ongoing use of certain material tools (marks, in the case of written genres) in certain ways that worked once and might work again, a typified, tool-mediated response to conditions recognized by participants as recurring" (p. 43). Spinuzzi (2003) highlights that genres may be addressed as a community's history of problem solving, "a sort of social memory that its practitioners accept without their explicit recognition that they are doing so," and that genres "provide us with ready-made strategies for interpreting not just discourse in a genre, but the world as seen through the "eyes" of that genre" (p. 43).

Consider an administrative or medical form, a personal homepage, an e-commerce website, a video streaming service, or an email inbox. We recognize these genres and, having recognized them, use the knowledge we have associated to each genre to interpret what we see and act accordingly. And we are surprised and/or lost when, for instance, the buying and paying process that the e-commerce pages implement differs from what we expect, what we expect being nothing other than a socially shaped process that we have integrated. In other words, the genre as perceived and/or understood by the user may differ from the one that drove the design and, to some extent, was implemented in the artifact (Spinuzzi, 2003).

Part of the answer to the question of how we perceive technologies and the way we can (or should) use them can thus be found in the genres that we perceive. Genres, as "traditions of using tools," convey a worldview and allow subjects to recognize the activity and the appropriate actions (Russell, 2009). The involved artifacts are associated with interpretative habits, and genre systems structure activities as both tacit habitual mechanisms and as devices (Spinuzzi & Zachry, 2000). Spinuzzi (2003) refers to this process as *genre perception*, which he defines as understanding a given artifact in terms of genre and applying one's knowledge of the genre to it.

Well in line with the perspective I develop in this chapter, the genre research tradition has evidenced that genre stems from how people's intentions and needs meet technologies' affordances rather than being a product of the technology's characteristics alone, including for innovative digital technologies. Miller and Shepherd's (2009) study of blogs well illustrates this point. These authors point out that "when [blogs] were new, the medium was the genre"—that is, there was an initial "blog genre." However, "adoption and experimentation [I would say: appropriation] led to differentiation and the multiplication of genres anchored in the same medium" (p. 284). "Journalism blogs," "team blogs," "photo blogs," "classroom blogs," and "travel blogs" all have their own characteristics. For instance, according to the authors, the specific "personal blog genre" stems from (1) the exigences related to "postmodern destabilization of the self," "endless play of subjectivity in a time of mediated voyeurism," and "challenges to the boundaries between public and private"; (2) the technology-specific affordances (e.g., means for immediate and reverse chronology editing or hyperlinks); and (3) the audience expectancies (e.g., frequent updating or authenticity). Similarly, email genres find their origins and characteristics in people's activities and not in the technology.

An important implication of the fact that genres mainly stem from individuals' or groups' needs and activities is that the "new" genres are impacted and shaped by how these previous needs were met, with or without artifacts such as paper documents or digital technologies. For instance, Spinuzzi (2003) has argued that dialog boxes are not paper-based forms but may be seen as continuing their tradition: "[They] are at least partially the same genre and imply the same worldview, the same understanding of the activity and what it values" (p. 42). As another example, Yates and Orlikowski (2007) showed that the business presentation genre preexisting the use of presentation applications impacted the initial PowerPoint genres. I will come back to the importance of considering the previous ways of doing of individuals and groups, and how approaches such as genre tracing (Spinuzzi, 2003) may help, in section 3.5.

From an ecology of artifacts perspective, it can be noted that genres may involve different media and artifacts. Moreover, as workers (and, more generally, users) coordinate clusters of artifacts to help them to get their jobs done (or, more generally, to address their tasks), what develops around activities is often a compound mediation involving a dynamic system of genres, which may be addressed as a genre ecology (Spinuzzi, 2003; Spinuzzi, 2001; Spinuzzi & Zachry, 2000; Yates & Orlikowski, 2002).

Stability and evolution   With respect to understanding the uses of technologies, an important finding is that genres are stabilized-for-now structures, which evolve. Previously used means may lead to the development of an initial genre that later gives way to more specific genres building on both the specific affordances of the new technology and the preexisting, conventionalized ways that have crystallized. These new genres present varying levels of traces of their "ancestors," which may lead to the development of new hybrid genres rather than a simple copy (Spinuzzi, 2003).

As it is fairly obvious that the initial email genre stemmed from letter-based communication let's take as another illustration selfies—that is, photos taken by, and including, the photographer. A specific use of smartphones is to take and exchange selfies to communicate our emotions or mood to friends. This specific use is now shaped by usual ways of doing and informal codes that young people know, recognize, use, and, sometimes, voluntarily transgress. Later on, new behaviors developed: indicating how one feels by sending "not-selfies"—that is, pictures of other persons, animals, or objects. This practice is based on, and participates in, the development of new technical means and resources such as repositories of funny pictures (Tiidenberg & Whelan, 2017). This emerging practice is likely to be impacted by the selfie genre the actors already recognize and enact, which may lead them to import (or transgress) its existing patterns or codes. Making sense of such evolution of practices raises questions such as (instantiating them on the selfie example): Does a specific not-selfie self-representation genre emerge? That is, is it recognized as such by the involved actors? Is there a kind of abstract self-representation genre in which sending selfies and not-selfies are two separate subcases, or are there related but different genres? Do the actors consciously or unconsciously make the connection? Answering such questions requires open-ended empirical inquiries into what patterns the actors recognize to be socially meaningful, and the answers may be local and temporary.

Another interesting finding on genre evolution is that genres are in constant negotiation. This has for instance been shown by Schryer et al.'s (2009) analysis of how "physicians, as the writers of child maltreatment forensic reports, are in the challenging position of having to suggest through their documentation that someone, often a caregiver, has injured a child through maltreatment without actually stating their conclusion" (p. 216). The analysis shows that these letters function as boundary objects because they are in a constant state of negotiation between their readers and their writers, and evolve to address emerging concerns. Referring to Bakhtin, the authors recall that "all verbal communication, whether written or spoken, is 'dialogic' in that to speak or write is always to reveal the influence of, refer to, or to take up in some way, what has been said/written before, and simultaneously to anticipate the responses of actual, potential or imagined readers/listeners" (p. 224).

Let's consider in this light what happens with digital technologies. When users' activities lead a genre (a conventionalized way of solving certain tasks or addressing certain goals) to evolve or develop, the very meaning of the involved artifacts is also modified. An email, an administration website, or an online discussion in the 2020s is not the same as in the 2010s or the 2000s. Designers can and do react to genre evolution and are also actors of this evolution. Taking a life-cycle perspective, it may thus be considered that the meaning of the application is subject to constant dialogical negotiation: The outputs of designers' work are taken up by users (appropriation, development of new uses), and the output of this process is in turn taken up and/or responded to by designers (technical evolution). I will come back to this in chapter 7. This macro social mechanism can be associated with more local negotiation mechanisms and outputs. For instance, emojis are generally thought to be used as kind of universal language for expressing emotions or concepts. However, studies show that groups (friends, family members, partners) also repurpose emojis and use them to express different meanings than those for which they were initially intended (Wiseman & Gould, 2018).

In a nutshell: How we perceive technologies (and, in particular, their usefulness) and the way we can (or should) use them involves both low-level psychological mechanisms (see the discussion on affordances) and high-level social forces (see the discussion on the structuration and genre approaches).

2.5 Synthesis

The points developed in this chapter may be synthesized as follows:

- Digital technologies are mediators. They must not be considered as something that people work on but as something through which people act.
- Usefulness is a key driver of both adoption and appropriation.
- With respect to appropriation, the usefulness in question is that perceived by users in light of their actual activities, which may differ from designers' expectations. It is subjective and possibly idiosyncratic. (And, although it may go without saying, understanding appropriation requires the consideration of user perception without any value-centered external judgments; understanding appropriation must not be confused with changing user practices, which may be a legitimate goal but one of a different nature).
- Different factors may shape and/or influence individuals' recognition and conceptualization of tasks and associated actions, and thus their actual activities. They include socially carried worldviews and ways of doing; artifacts, whose raison d'être and design implement and convey previous experiences; and individuals' conceptualizations, knowledge, beliefs, skills, experience, expectations, or perception and meaning-making mechanisms (i.e., the developed psychological constructions).
- Perceptions and understandings both crystallize (i.e., stabilize temporarily) and evolve: People (and functional organs) develop.
- Perception of technologies usefulness is related to their effecter affordances (i.e., the possibilities of the artifact for making an effect on an object) rather than their handling affordances (i.e., the possibilities for interacting with the artifact).
- People may and generally are engaged in simultaneous activities, including high-level and possibility unconscious activities, and interplay between these activities may affect the uses of the different involved technologies.
- Understanding the uses of a given technology requires the consideration of the user technical ecosystem and its dynamics: user ecology of artifacts, spaces/places, and instrument-feature and/or instrument-artifact relations (which may not be 1–1 relations).

In the three following chapters, I will propose an understanding of appropriation that is based on the aforementioned points: Chapter 3 describes a theoretical account of appropriation, chapter 4 complements it by studying the developmental aspects in closer detail, and chapter 5 further explores the high-level and possibly unconscious activities that may play a role.

Before I do so, the following final section returns to the focus on the individual that is adopted in this book and was sketched out in chapter 1, and which can now be discussed in closer detail.

2.6 Addressing the Social and the Individual Levels

Acknowledging the Social Dimension . . .

As we saw in this chapter, social forces are much more than a factor affecting the initial technology adoption: They convey more or less explicit patterned ways of addressing tasks and/or using technologies. Via the way they are embedded in technologies and/or accompany them, these patterns play a role in how we recognize the activity and the appropriate actions: They shape our beliefs and/or expectations and also define what is valued and the different roles to be played. Social forces thus influence and sometimes structure how we perceive technologies and engage in action.

In particular, collaborative work practices and professions play a key role in the case of workers, and the analytical approaches we reviewed in this chapter can be articulated around this specific point. Activity theory–oriented studies of workplaces or organizations have shown that individuals' activities could not be understood outside of their social (and historical and cultural) context. The classic means to achieve this is Engeström's (1987) activity system model, which extends AT original subject-object-mediating artifacts triangle with notions such as the division of labor (division of activities among the actors) or the rules (e.g., conventions or guidelines) regulating the interactions between actors. As I have already emphasized the genre approach, I will simply stress the congruence with AT. Engeström (2009) points out that activity and genres (as "classifications of artifacts-plus-intentions") may be seen as complementary units of analysis. Russell (2009) notes that "much of Engeström's research, like [the genre approach], takes organizations as its primary research object and attempts to explain change and stability in these dynamic contexts historically and developmentally" (p. 51). Spinuzzi (2003) highlights how genres can be addressed, coherently with the AT perspective, at different levels: as social action ("material residue of a community's problem solving as it engages in an activity [embedding] community values and beliefs" [p. 127]); as tools, strategies or tactics ("the implements used to perform the many goal-directed actions in which community members are engaged" [p. 136]); and as unconscious collections of operations ("learned habits or responses on which a worker unconsciously draws" [p. 144]). An example of the latter is how a user may "transparently" fill a dialog box without thinking about the interface.

The role of work practices has also been highlighted by other research traditions, for instance the ergonomics tradition centered on human activity in the workplace, which in many respects echoes the AT perspective on activity (Daniellou & Rabardel, 2005). As mentioned by these authors, "Activity is not only a relationship between a subject and an object. It is also a relation with other subjects, who may be physically present or present via instruments and tools, sign systems and rules and procedures that mediate activity" (p. 355), and "what takes place in the activity is not always what is described in the procedures and reference manuals produced by management, designers or organizers" (p. 356). Focusing on the preservation and development of operators' power, Clot and Faïta (2000) see genres as social prescriptions—that is, as the tacit part of the activity that is known and shared by the workers of a given setting and defines what they are expected to do without having to respecify tasks each time they appear.

As a final example, the occupational perspective on the technology choices of design engineers' work conducted by Bailey and Leonardi (2015) reveals how their sociotechnical arrangements differed in nature despite the fact that they conducted fundamentally similar tasks (e.g., sketching, calculating, modeling, and testing) and all used computers. In the words of the authors, "Individuals' ideas and beliefs at work with respect to the technology choices that face them derive from occupational factors" (p. 197).

We saw how the role of social mechanisms is also very obvious in the inception and evolution of social media technologies (e.g., blogs, wikis, discussion forums, or online review sites) and social networks. Another good example is online games: How the setting should be interpreted, what the gamers are expected to do, and how they should do so are socially conveyed.

. . . Must Not Overshadow What Happens at the Level of the Individual

However, Alice and Bob (to use names that computer scientists often use for individuals) are not just instances of abstract entities (e.g., "engineer" or "young people") using a type of application (e.g., collaboration platforms or social networks) or even a specific application (e.g., Slack or Facebook). Alice and Bob are first and foremost individuals with their own history, motivations, values, ecology of artifacts, ways of doing, perceptions, meaning-making mechanisms, conceptualizations, or, as we will see in chapter 5, emotional or existential issues. All these aspects play a role in why and how they appropriate technologies and if and how these technologies mediate their activities in a way that is efficient and coherent with their values.

The fact that Alice and Bob are individuals does not mean that their activity is purely individual or even individual at all. As we saw in this chapter, activity is always social, although this social dimension may be conveyed by habits, implicit rules, structures, or designs and not simply by the actors present. When we engage in tasks such as "getting my job done" or "gaining social capital," we (often unreflectively) use cultural tools, and the technologies we use (e.g., office suites, collaborative platforms, or communication apps) do reflect cultural and social constructions. More generally, Alice's and Bob's activities include acting, as individuals, on individual objects; acting, as individuals, on collective objects; and acting on collective objects as members of collectives such as the engineering department where Alice works or Bob's group of friends. Whatever the individual or collective nature of the acting entity and of the object of the activity, Alice and Bob conduct their activities in a social way.

Nevertheless, and as recognized by all activity-oriented perspectives, "activity is always unique. It is specific to given subjects in a given context. . . . The same task carried out by people with different characteristics will not generate the same activity. . . . Activity cannot be understood only in terms of normal tasks prescribed by management or by designers [it] is a specific construction, in response to a specific context. . . . [It] is affected by the subject's life experience and is, thus, constantly revised and reinvested" (Daniellou & Rabardel, 2005, p. 355).

Moreover, the conceptualizations and psychological constructions that underlie and shape our perception and meaning-making mechanisms are not social. They are ours (although, once again, we do not develop them in a vacuum and the social dimension is core). What acts as a medium, and how it does so, thus differs from one individual to another.

Finally, most of the everyday technologies we use are fairly general. We use our communication and social networking apps, word processors, or spreadsheets for a variety of tasks. When the "aligning numbers" technique elaborated in the history of mathematics has few uses other than multiplication, most of the digital technologies we use offer much more space for individuals' characteristics or life experience to play a role, and for innovative uses to develop.

The social dimension of appropriation should not therefore overshadow the individual (idiosyncratic) aspects, and this is why this book mainly considers users as individuals. Although cultural and social constructions do shape many of our practices and influence most if not all of them, our practices are not necessary consequences of these constructions. Meaning is something that is personal and subjective; it is influenced but not imposed by structures (Mekler & Hornbæk, 2019). Moreover, we can also decide to use technologies (or to use them in one way or another) according to our personal values; see the notion of "existential users" in chapter 5.

Acknowledging these hopeful characteristics is crucial to both understanding appropriation and designing technologies that empower people. Alice and Bob are entitled to understand what they do and how they do so as real people rather than simply as generic profiles.

It is fair to say that some of the works building from structural, social, or genre analyses do mention that care must be taken to avoid putting too much emphasis on the institutional aspects and the evolution of social practices. de Sanctis and Poole (1994) mentioned that the role of the organizational environment or the users' collective functioning should not lead to an underestimation of the role played by the properties of technology. Spinuzzi (2001) highlighted that although genres do impact and often shape uses, there is still a possibility that idiosyncratic divergent understandings and uses of artifacts may develop within a given cultural-historical milieu. Orlikowski (2000) underlined that what is core is how people mobilize the social structures embodied in the technology.

Although these points make sense, an understanding of how individuals appropriate technologies as individuals (and not solely as members of a collective), and of the role of idiosyncratic and personal dimensions (e.g., the development of psychological constructions), is absolutely essential. This is why one of the specificities of this book is to propose a perspective that natively focuses on the individual.

3 An Activity-Centric Theoretical Account

This chapter articulates and further develops the analyses conducted so far to propose an integrated understanding of appropriation. I successively introduce and illustrate the notion of *functional value*, which captures how the notion of functional organs resonates with users' perceptions and meaning-making mechanisms (section 3.1); show how this notion helps to make sense of uses (section 3.2); propose a general analytical framework for appropriation that builds on this notion (section 3.3); introduce the notion of *functional transparency* as an important characteristic of appropriation (section 3.4); and finally summarize the questions addressed and the questions raised (section 3.5).

3.1 From Functional Organs to Functional Values

Functional Values

We saw in chapter 2 that the development of functional organs—that is, functionally integrated and goal-oriented configurations of internal and external resources—pertinently qualifies the outputs of an appropriation process: The technology has been associated with preexisting and/or new ways of doing and included in the individual's repertoire of resources. From a mediation and activity perspective, it has become an amplifier of human capacities.

In order to step from a characterization of the output to an understanding of the process, we now need to consider what drives the development of functional organs. The analyses developed in chapter 2 provide us with (1) two important principles, namely that the distribution of activities between mind and artifacts is always functional (Kaptelinin & Nardi, 2006) and this functional dimension must be addressed from the perspective of the user, and (2) a number of important points listed in section 2.5. These include the driving role of perceived usefulness as redefined in section 2.1—that is, the subjective and possibly idiosyncratic usefulness that is perceived by users in the light of their actual activities.

A way to simultaneously address the functional and perceptional dimensions of the user-artifact link is to consider the notion of functional value using the following definition: "The functional value of an artifact for an actor is the utility of this artifact for achieving some tasks or goals as perceived by this actor" (Tchounikine, 2017, p. 159). What is at play with functional values is user's perspective on the user-artifact relationship: Users consider certain tasks and attribute functional values to digital technologies when, in practice, they perceive them as a means to achieve the tasks they consider.

The rationale for introducing this functional value notion is that it is instrumental to (1) capture the functional and perceptional dimensions of the user-artifact link and (2) draw attention to users' actual motivations. Coherently with the argument developed in section 2.1, the perception considered here is the users' perception and understanding of their actual practices.

Illustrations

Let's first consider some of the examples used in preceding chapters.

The way the email example was introduced in chapter 1 may be reframed as follows. Some people attribute a "to-do list" functional value to their inbox. As an output, the email inbox becomes part of the integrated set of resources for remembering things to be done or, in other words, extends the individual's remembering capacities.

As a counterexample, it could be said that many people attribute a general "communication" functional value to social networking applications. However, this statement is too general to be instrumental. Nobody communicates without a purpose. What is at stake is thus rather an interplay of more specific functional values such as "maintaining/developing social life" or "gaining social capital". Social networking applications extend their users' socializing capacities.

Let's consider the use of collaborative platforms such as Slack in the same detail. The rationale for the platform and its spirit—that is, communicating via project-, task-, or role-specific channels—is to promote and support collaboration. However, in examples such as that reported by Stray and Moe (2020), some individuals may not use these channels and, despite the social and institutional pressure, prefer to use the platform for one-to-one communication means such as personal messaging or email. This could be interpreted as follows: Some workers appropriated the platform when some others did not. This perspective, however, implicitly considers the expected way of using the platform or, in other words, a faithful appropriation. Taking a functional value perspective, I would rather argue that some workers attribute a "one-to-one communication" functional value to the platform and that this value is given priority over that attributed to channels, if the latter exists. (This pattern is fairly frequent. It may stem from past workplace hierarchical communication structures and/or, as suggested by current knowledge about programmers' psychology, other factors such as avoiding social pressure, maintaining personal alliances, or preserving authorship. As we will see, functional values may relate to symbolic or psychological ends.)

The examples used when studying the affordance notion can be similarly reframed: The screw or the piece of wood used to push the cork inside the bottle are given a (one-off) "remove the cork" functional value; the toolbox and the smartphone are given an "offering possible means" functional value; and so on. Here, the clear priority is the possibility of having an effect and not how the system should be handled. A screw suggests both handling and effect affordances similar to that of a corkscrew, but this is not the case for the piece of wood used to push the cork inside the bottle. Similarly, nothing in the undo-redo buttons of word processors suggests a potential use to find where we were in the text by clicking once on each button.

An IoT button that is configured to control the smart home "close the shutters" feature and that is placed on the way to the bedroom is given (1) a "close the shutters" functional value (which is related to the associated feature) and (2) a "remember to close the shutters" functional value (which is related to its location).

Let's now consider new examples.

Smartwatches offer different features such as communication means, GPS positioning, health trackers, and, of course, displaying the time. As for all technologies introducing new technical features, a number of works have studied what features were the most frequently used on discovering the device, over the long run, or according to people's profiles. More interestingly, Pizza et al.'s (2016) empirical study has examined *how* these devices were used in actual practices, revealing several specific functional values. One of these is notification triage. Rather than responding to incoming messages, which is usually done via the smartphone, the watch is used to identify the need for a quick reply without physical disengagement from the current activity. A positive by-product functional value is to provide time to think about the incoming message and respond more quickly or accurately later on. Another positive by-product functional value is a reduced need, or temptation, to take out the smartphone and be impolite. As a side remark, the combined uses of the smartwatch and the smartphone that are identified in Pizza et al.'s study once again illustrate the importance of taking an ecology of artifacts perspective.

Smart speakers combine voice recognition and smart assistance features to allow their users to control other devices or ask for services such as playing music, telling jokes, answering basic questions, managing specific actions (e.g., adding an item to a shopping or to-do list), or reminding the user that it is time to do something. Here again, use-oriented studies allow us to identify precise functional values, for instance in parenting (Beneteau et al., 2020). One of these is making children practice speech and language skills. Another is making the device (Alexa) say that it's time to stop playing and get ready for bed. What is specifically interesting about this unexpected use is that the central issue is not just the time management feature. The parents indicate that the use of other technologies (e.g., their smartphone or a basic timer) would not be perceived in the same way. As put forward by the authors, here we see a neutral third-party mediator functional value, which leads to specific uses of different artifacts (in the words of the authors: "The parent clarified that they had surreptitiously set the timer with the app on their phone so the children would perceive the timer being generated by Alexa, rather than the parent" [p. 7]).

As a final example, Griggio et al. (2019) found that, for people who use multiple communication apps and adapt them, customizing one app influences the customization of others. Coherently with other examples we saw such as the repurposing of emojis or the role of social aspects in meaning-making mechanisms (see section 2.4), the authors found that "participants tailored their apps to express their identities, organizational culture, and intimate bonds with others." As a side-effect, the users experienced "frustrations around barriers to transferring personal forms of expression across apps" (p. 1). This can be understood by considering that users attribute idiosyncratic expression-related functional values to the apps and, in this context, adapt them, thus increasing the strength and the specificity of the user-functional value-app link. When they change these apps, they search for a way to continue benefiting from these functional values and therefore have to adapt the new apps accordingly. This echoes the "making new software look like the old one" general phenomenon.

As illustrated by these examples, applying a functional value perspective to the use of technologies leads us to consider users as actors, to focus on their activities, and to identify their actual and possibly contextual motivations and perceptions. It thus drives the analysis in a way that is coherent with the points made in chapters 1 and 2.

3.2 Making Sense of Uses by Focusing on Functional Values

Let's now go into greater detail and consider two important points, which I will artificially dissociate for the sake of explanation: the importance of considering precise functional values and where these values originate from. The studies taken as illustrations in this section address the use of technologies by teachers, a domain that has been the subject of detailed analyses.

Considering Precise Functional Values

Among the numerous analyses of the important characteristics of technologies for teaching and learning, several studies have examined if and how teachers used social networks and communication apps to interact with their students. Here, I will focus specifically on what teachers seek to achieve via this strategy.

Asterhan and Rosenberg (2015) analyses of Facebook by secondary school teachers revealed that their self-reported motivations include very specific and precise goals. Examples include monitoring and appraising the students' world, improving personal relations with students, improving teacher status (e.g., "Show them different sides of my personality" or "Raise my status"), or academic purposes such as triggering discussions or managing homework. A similar analysis of WhatsApp uses (Rosenberg & Asterhan, 2018) reveals how, in my words, some teachers attribute functional values to the technology such as updating students about the school's ongoing activities or enforcing discipline.

The analyses of the students' uses also reveal very specific and subtle functional values. One of these is avoidant communication. In the studied WhatsApp students+teacher group, a message from a student is interpreted as a collective message from the entire group, while a personal and private message to the teacher is a means to communicate as an individual (Rosenberg & Asterhan, 2018). This avoidant communication functional value turns the technology into a means to raise issues that would not be discussed in face-to-face situations. It may be noted that the functional value is attributed to the group forum and not to the application. Actually, the application is also given other functional values. For instance, there is a student-only group, which serves other needs.

As one can see from these examples, understanding uses requires considering *precise* functional values. It is not incorrect to synthesize such analyses as evidence that teachers attributed a general "communication with student" or a "pedagogical" functional value to Facebook or WhatsApp, and this level of analysis is sufficient to explain why some teachers adopt these technologies. However, the uses that are developed are related to the teachers' precise activities. Showing a new side of one's personality, provoking discussions, and enforcing discipline are activities that differ in nature. Technological features and/or associated social mechanisms may align with how teachers intend to reach some of these goals but complicate the achievement of others (see the next section).

A second important point is that the goals that teachers self-report are not specific to the use of Facebook or WhatsApp, or indeed to digital technologies. They are pedagogical goals that teachers already addressed before using apps and continue to address in class with other means. The functional value that is attributed to the technology is thus to be regarded within this holistic perspective.

What surfaces here is how, in line with the analyses developed in chapter 2, the functional values attributed to the technology originate from the teachers' usual practices, and not from the technology. Let's develop this point, using a more precise example and illustrating how such a process may create issues with the existing technological features.

Focusing on Users and Their Practices

A learning scenario editor is a system that offers teachers an interface to model the tasks that a group of students are expected to perform. Such editors feature notions such as students, groups, roles, activities or resources, and a temporal organization. The main goal is to support teachers in reflecting on and refining their scenarios, a concern that is at the heart of their teaching activities.

The tests of the learning scenario editor designed by Sobreira and Tchounikine (2012, 2015) revealed how, although all teachers perceived the system as a means for editing the general structure of classroom activities—that is, what the system is designed for—most of them also developed other perceptions. For instance, K–12 teachers mentioned perceived functional values such as having a clear record of what was done in the classroom or sharing pedagogic constructions with colleagues, and student-teachers in pre-service education mentioned functional values such as checking that student activities planned as parallel sessions would not require the teacher's presence in different groups at the same time.

What is specifically interesting here is the nature of these functional values. The K–12 teachers were all experienced teachers. Having a clear record of what was done in the classroom as a resource to be annotated and reused for future sessions, or sharing pedagogical constructions with colleagues, are functional values that stem from their work practices and, more precisely, how they continuously improve these practices. The editor is not designed for such purposes. It is the fact that these users have these concerns that leads them to perceive these values. Similarly, as checking the coherence of the classroom preparation is an issue inexperienced teachers find hard to manage, this leads them to perceive the editor as a way to achieve constraint-checking tasks.

The importance of considering precise functional values is well illustrated by how some K–12 teachers unexpectedly considered the system as a means to produce clear proof of classroom preparation, which resonates with the demonstrability factor (i.e., if and how results of adopting the technology is observable and communicable to others) highlighted by some of the TAM models. Teachers who only or mainly attribute such a functional value to the editor may construct scenarios via other means and use the system to represent them only. In such cases, the objective of supporting the elaboration of the scenario would not be met. Similarly, when students are offered digital tools designed to scaffold learning but use them in unexpected ways, the teaching objective is not met (Tchounikine, 2016, 2011). Actually, from both an appropriation and an emancipation perspective, learners should be empowered to select the applications they use and, more generally, the support they obtain from these technologies (Tchounikine, 2019b).

Important Points When Analyzing Functional Values

The examples studied in this section illustrate a series of important points.

First, understanding why and how individuals appropriate technologies requires the consideration of *precise* and possibly *idiosyncratic* functional values. Analyses of uses in the light of predefined general categories (e.g., efficiency, play, excellence, aesthetic, status, ethics or esteem) may inform large-scale analyses and design considerations; see Benamar et al.'s (2020) general analysis of IoT devices appropriation as an example. However, such top-down analyses blur the precise interests and semantics that users attribute to the technology.

The second point, which we have already discussed in chapter 2, is that focusing on precise functional values requires a *holistic* and *multiple-level* analysis of users' *actual motivations and activities*. Typically, the goal of teachers is not to use Facebook or WhatsApp or a scenario editor; their goal is to solve different educational problems related to preparing the class and managing students. To understand how teachers appropriate the technologies they have at their disposal, it is thus important to identify these educational problems, consider them as such, and examine how the teachers address them (as opposed to focusing on what the new technology reveals). Similarly, the goal of software developers is not to use Slack to collaborate but to get their programming job done (and/or, possibly, promote themselves), and the goal of young people is not to use Instagram but to serve precise socializing ends. To understand how developers or young people use these technologies and appropriate them, we must therefore first and foremost understand their activities and the factors influencing them—for example, their work practices or social practices (see chapter 2).

Third, attention must be drawn to the fact that people (and, in particular, workers) generally already use a specific way to address the goals that new technologies are supposed to support. They have developed experience and idiosyncratic practices. Within this context, the perceived or attributed-in-practice functional values of a new application are related to how the system may help them to achieve their tasks *given their preexisting personal representations and work practices*. Practices evolve, and new technologies play a role in this evolution. However, providing people with a new device does generally not suffice to make them develop a particular conceptualization or practice.

Finally, the functional values attributed to a new technology are related to those attributed to other technologies. This echoes the general perspective according to which the use of a new system must be regarded as a modification of the ecology of artifacts that people use to mediate their activities.

3.3   A General Analytical Framework for Appropriation

Let's now see how the notion of functional value enables us to link together the points made in chapter 2 and, more specifically, consider the mediation perspective (section 2.3) and the psychological dimensions of perception and meaning-making mechanisms (section 2.4) in a way that focuses on the individual while acknowledging the role of social forces (section 2.3 and discussion in section 2.6).

Users Interact with Tasks and, in This Context, Create Resources

To explain the perspective I propose, I will contrast it with two prototypical approaches.

The basic approach to the study of human–computer interaction, which underlies some of the naive approaches to the appropriation phenomenon, is to consider how users interact with applications—for example, how they use email clients (see figure 3.1a).

In contrast, the traditional mediation approach suggests considering how people act through technologies (see figure 3.1b). With regard to the understanding of appropriation, this makes much more sense. As shown by the studies reported in chapter 2 and clearly highlighted when taking a functional value perspective (see sections 3.1 and 3.2), appropriation stems from the actual activities of users. Technologies are just potential media with characteristics that play an important role but, nevertheless, remain media and are not drivers. When attempting to understand appropriation, focusing on how people interact with applications—that is, considering the technology as something that people work on—is thus misleading. Taking a mediation approach pertinently suggests that the analytical focus should rather be on how actors interact with tasks.

The analyses I developed in chapter 2, which highlight the developmental dimension of appropriation, lead me to propose a slightly different perspective (see figure 3.1c): When considering appropriation (and not solely use), the analysis-driving axis must be that users interact with tasks and, in this context, create resources (remember that functional organs are "functionally integrated, goal-oriented configurations of internal and external

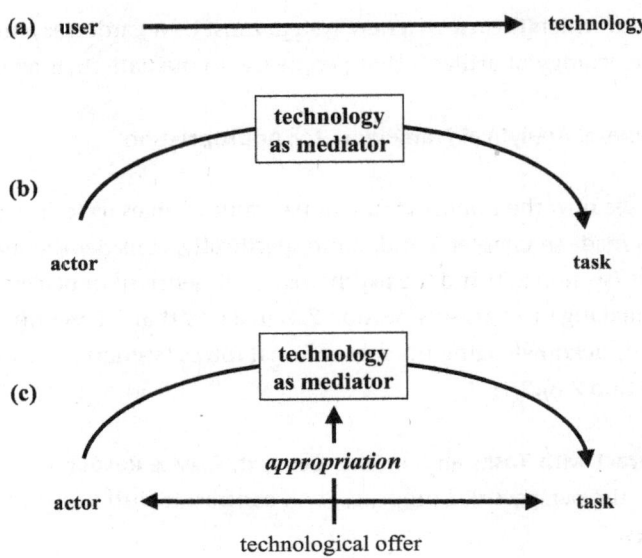

Figure 3.1
Perspectives on technologies mediation and appropriation.

resources," and the internal resources, namely psychological constructions such as notions or mental schemes, are developed *by the user*). What mediates user activity is not therefore the technology as it was initially designed, but the resource that users have developed for themselves—that is, the outputs of their appropriation of the technological offer. This includes the involved technologies and the different social and psychological constructions related to the individual–technology and individual–tasks relations.

Artifacts Become Mediators/Resources When Given a Functional Value

Taking this perspective, an artifact becomes a resource (referring to figure 3.1c, an item of the technological offer becomes a mediator) when the user has attributed one or several functional values to it—that is, a utility for achieving some task as perceived and addressed by him/her, and has associated it with psychological constructions (either preexisting or developed on this occasion) that allow/drive its use for these tasks. This takes place within the social context (e.g., the work practice) and the stable-for-now actor–task relations developed so far.

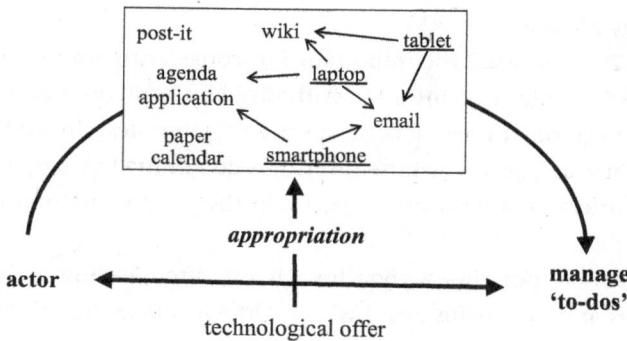

Figure 3.2
Taking an ecology of artifacts perspective to mediation and appropriation.

Moreover, as a direct implication of the point stating that it is essential to consider user ecology of artifact and its organization (see section 2.2), the artifact–functional value relation is not to be regarded as a 1:1 relation. Users' interaction with tasks is generally not limited to the use of one application or, more generally, one resource. As we saw with our running to-do-list example, people generally use resources that are both digital (e.g., electronic agenda, notepad, wiki, or email applications) and nondigital (e.g., paper-based agendas or Post-its). The digital resources may be accessible via different artifacts (laptops, tablets, or smartphones), which may partly explain why they are attributed a functional value. Artifacts may also include devices that are contingently used for implementing crystallized high-level strategies—for example, something placed next to the glasses case or the car keys. One artifact may be attributed several functional values: This is very obvious when considering hosts (e.g., smartphones or smart speakers) or general-purpose means such as communication apps, but it also holds for technologies designed as task-specific (see the example of the learning scenario editor). Finally, functional values associated to one application may be related to those associated to other applications, and/or stem from the coordinated use of different artifacts (see for example the coordinated use of smartwatches and smartphones). Figure 3.2 proposes a version of figure 3.1c featuring this ecology of artifacts perspective.

Before we step forward and study the connection between our motives and the means in more detail, let's develop the ecology metaphor.

The Ecology Metaphor

In section 2.2, I stressed the importance of considering user ecologies of artifacts as dynamic structures. As synthesized by Kaptelinin and Bannon (2012), the ecology of artifacts perspective "[promotes] an understanding of artifacts as 'species,' whose survival is determined by a dynamically unfolding interaction with other species in their shared natural environment" (p. 290).

Metaphors are not always good for science, often leading to erroneous conceptions and/or conclusions. I will nevertheless take the risk of further developing this ecology metaphor while keeping my focus on people, the functional values they perceive, and the psychological constructions they develop.

Taking an abstract perspective, it may be said that the members of our ecology of artifacts are involved in a form of coevolution. When designers design new devices, new apps, and/or new versions of these apps, they take into consideration the current technical offer and its uses. If we take a step back and consider as "species" the different everyday technologies we are offered to use (e.g., Web technologies and communication apps), we can see that they are increasingly interrelated (they technically communicate), and both their breakdowns and successes lead to immediate reactions (e.g., new apps or features taking advantage of the newly opened possibilities).

As users, we participate in this biological-like technical effervescence. We inspire designers via our innovative uses and articulations of technologies (see for instance how "the Web" as we experience it now developed via innovative usages of an initially basic network service), we resolve the issues resulting from new versions by adapting our uses and, sometimes, adapt the technologies (see chapter 6).

Taking a functional value perspective helps us to identify and make sense of the relationships and interactions that underlie part of this effervescence. A basic example is how email and agenda apps have coevolved. As, in this case, the major reasons for this coevolution (e.g., the role of email in organizations) stem from general structural forces and thus apply to many if not most users, the coevolution is highly obvious, and taking a functional value perspective only adds a theoretical explanation. However, as illustrated in chapter 2 and sections 3.1 and 3.2, individuals also relate artifacts to each other for less obvious, odd, and/or purely idiosyncratic reasons that can only be revealed by taking an explicit functional value perspective.

Moreover, another dimension of this biological-like effervescence is that, as I will study in closer detail in chapter 4, we (the human users) are part of the coadaptation and coevolution process: We literally adapt our mental structures and, also, develop new ones.

3.4   From Use Transparency to Functional Transparency

When taking a functional value analytical perspective, the way research traditions such as AT or the genre approach shed light on how we make the connection between our motives and the means we consider (see section 2.4) leads to the following observation: While use transparency may play a role, appropriation is rather related to *functional transparency*.

Use Transparency

As we saw in section 2.3, use transparency refers to the state where users addressing a task via an application think about the task, and not about the system they are using. In other words, the mediator is invisible. Users do of course see the system interface, but its use does not generate interfering conscious processes (Kaptelinin & Nardi, 2006).

It is now widely accepted by even the more technology-centered approaches that "transparency is not a static feature of the interface but an evolving qualitative aspect of the relation between human beings and computer artefacts" (Bardram & Bertelsen, 1995, p. 89). It is "a quality of the use activity [that] is developed by the user during interaction" (p. 79).

Taking an AT perspective, transparency is an output of automatization— that is, the transformation of actions into operations. The first time we use an app (or drive a car) we have to consider in an explicit way how some actions (e.g., editing and sending a new email or starting the car) must be carried out through operations (e.g., by selecting a menu option or turning a key). Later on, we no longer need to think about how to carry out these actions; they become operations and are performed without conscious thought.

The transformation of actions into operations is directly related to use frequency and is not to be confused with an inflexible behavioral pattern. It takes place within more general structures, namely the system of activities in which the user is engaged, and there is a contextual dimension. At any moment, attention may be drawn back to how actions should be achieved and/or, coming back to the activity level, what actions should be considered.

The role of attention is well illustrated by Rochat et al.'s (2019) study of runners' sensorimotor appropriation of devices such as water bags, which they carry out from an enaction perspective. Although the sensorimotor phenomena involved are of course very different from what happens in the digital world, the analytical frame of mind resonates well with the points made above. Appropriation is used in this study to refer to when the artifact disappears from the runner's consciousness and is literally considered by him/her as an extension of his/her body. An interesting point, which once again illustrates the contextual dimension of users' interaction with tasks, is that in the runners' experience, the presence of the artifact (of the water bag) is not a binary present/absent static feature. The transparency of the artifact is induced by both its structural properties (weight, ties) and the actor's (the runner's) experience, this experience being impacted by the characteristics of the setting and the associated physical and cognitive activity—for example, the degree to which the portion of the trail is uneven and difficult to run and the consequent attention paid to avoiding falls. Using the authors' words, the reasons for this include that "cognition fundamentally depends on bodily processes (i.e., perceptions, actions) that are continuously regulated by inseparable couplings between the perception of the environment and the actor's behavior" (Rochat et al., 2019, p. 475).

The notion of use transparency thus well depicts a state of affairs that is directly related to appropriation: When we have appropriated a technology, its use generally ceases to generate any interfering conscious processes. Conversely, when the use of the system does generate issues, this may hinder appropriation.

However, returning to the importance of the functional dimension of the user-artifact link, use transparency fails to address a related, though different phenomenon: the mobilization of the means. I propose to address the latter by considering the notion of functional transparency.

Functional Transparency

Although use transparency is indeed part of the picture, what seems to better characterize appropriation is what I propose to label *functional transparency*—that is, the transparency of the task–application association as mediated by the functional value(s). Functional transparency relates to how a user immediately and transparently mobilizes a given set of resources, without any conscious explicit effort, when confronted with a task.

The essence of functional transparency is obvious when it does not occur. Returning to some of the examples used in the preceding chapters, appropriation is well illustrated by the straightforwardness of behaviors such as: need to take a note / open the pad; need to say something to friends / open WhatsApp; need to remember something / send an email to myself; and, in a nondigital example, need to open the bottle of wine / take the corkscrew from the kitchen drawer. However, when the application is not accessible—that is, the particular item of the configuration of internal and external resources developed for the task is not available—we have to explicitly consider the question of how the task could be addressed.

How we remember things to be done is a good illustration of the use/functional transparency dichotomy. What is core is the fact that, when facing a "remember to do something" task, we straightforwardly (transparently, without any conscious or explicit effort) mobilize one of the instruments (one of the technologies we have appropriated for this task) such as a pencil and a piece of paper, a note-taking app on the smartphone, an email to oneself, or an object put in an odd place. The fact that the actions for achieving this are carried out transparently is another question that, although related, is arguably different and partly contingent. Use transparency of course likely plays a role in the fact that, for instance, the inbox became a to-do-list instrument, and, conversely, using the inbox as a to-do list contributes to the fluency in system feature use. Nevertheless, the "to-do list" value of the inbox is not dependent on the email client use transparency. Indeed, as we saw, many users continue to use the inbox as a to-do list while experiencing difficulties with the management of these emails and being aware of these difficulties. The functional value withstands the use issues. Conversely, an online tax declaration may have become functionally transparent (when we have to report something, our first move is to sign in to the website, rather than visit the tax office or print and send paper documents as we did in the past), but the use is not transparent (because we only sign in to the site a few times a year, the declaration is complex, or we are cautious and/or stressed).

In some sense, the alignment of use transparency and functional transparency is a kind of ideal prototypical case, and when the application is used in an efficient way to tackle the task at hand, this is the icing on the cake. However, there may be functional transparency yet use opaqueness (use difficulties) and, vice versa, use transparency and functional opaqueness (although the application would allow us to achieve the task and is fluently used for other purposes, it is not immediately and transparently thought to).

Patterned Ways of Doing and One-Off Appropriations

In addition to clarifying the contingent nature of use transparency, the notion of functional transparency sheds light on another aspect of how we make the connection between our motives and the means we consider, namely one-off appropriations. What I call a *one-off appropriation* is the contextual use of a new device or feature to address a task.

One-off appropriations may be addressed as (1) an existing functionally transparent association of the task with a set of resources and (2) given the unavailability of the artifact one expected to use, the contextual search for new functional links. This search is shaped by the usual resources' characteristics including, in particular, the associated patterned ways of doing (patterned ways of addressing tasks and/or using technologies). Such new functional links may be ephemeral or become part of the stable, though evolutive, configuration of internal and external resources for the task at stake.

Let's return to the example of opening a bottle of wine in chapter 2. People who have appropriated the corkscrew technology transparently mobilize it. If a corkscrew is not available, they have to consciously solve the contextual problem of linking the task to another resource. As we saw, this may be addressed by opening the kitchen drawer or the toolbox and hoping that their attention will be drawn by a suitable effecter affordance—that is, an artifact that will have a suitable effect on the cork. This suitability is, consciously or unconsciously, partly built from the existing resource. The crystallized usual way of doing is a driver of what will be perceived as a potential means. This does exclude other perception mechanisms, such as having one's attention drawn by a device that has only a visual resemblance to the usual device used or triggers a different perspective on the problem to be solved.

Let's now consider the remembering-things-to-be-done example. The impossibility to send a to-do email to oneself may lead the user to resort to a digital means such as browsing the apps to detect a potential note-taking means, or a nondigital solution like putting a piece of paper or an object in a place where it will be seen. These means partly derive from the existing high-level strategy underlying the appropriation of email as a to-do list— that is, creating a situation that attracts our attention. While this is straightforward with the usual means, a new implementation of this strategy can require conscious efforts, for example to contextually solve the problem of deciding what object will be used and where it should be placed (see the previous discussion on actions and operations).

One-off appropriations may thus be seen as use cases within which users mobilize internal resources attached to other means. This perspective addresses the apparent contradiction between (1) the fact that appropriation builds on long time-span processes such as developing conceptualizations or usual ways of doing and (2) the here and now nature of one-off appropriations. Using a wood screw to open a bottle of wine or putting the saltshaker on the way to the bedroom has a one-off dimension but, also, a recurrent dimension (the way of doing), the latter having been shaped through the appropriation of other devices. We will come back to this in chapter 4.

3.5 Questions Addressed and Questions Raised

Building on the propositions presented in this chapter and in Tchounikine (2017), the issue of stating a precise definition of appropriation may be given the following response: Appropriation is the process by which users, while interacting with the tasks they consider, attribute functional values to digital artifacts and associate them with one or more patterned way(s) of (1) addressing the task at hand and (2) using the technology; the outputs include integration in practices and functional transparency.

This definition provides criteria for considering if and, more importantly, for what purpose, a device, application, or feature has been appropriated: One or several functional values have been attributed to it; it is associated with ways of doing, which may be imported or adapted from preexisting practices and/or develop in the context of the emergence of a functional value; it has become functionally transparent—that is, the artifact and its associated psychological constructions are immediately and without any conscious explicit effort mobilized by the user when he/she considers the tasks for which this artifact has been given a functional value. These criteria correspond to a stage of use where the user has established a stable-for-now artifact–task relation, and the artifact (and the ways of doing associated to the couple task-artifact) has become part of his/her configuration of internal and external resources. The associated functional transparency may be (and indeed often is) accompanied by use transparency.

The functional values that should be examined are those that crystallize in actual working or everyday use. During tests or preliminary uses, users may foresee functional values—that is, imagine the utility of a feature, application, or device for achieving some tasks or goals. Considered in this way, the notion of functional value rejoins TAM's notion of perceived usefulness, with the important difference that it relates to precise tasks/goals and not to the global usefulness of the technology. This dimension, which drives adoption and initial use, is important. Nevertheless, what is core to appropriation is the actual activity that develops in everyday uses and the functional values that emerge and/or crystallize in this context. These values may correspond, relate to, or have nothing to do with those the designers and/or the user had anticipated, and the usefulness in question is that subjectively perceived by users. As a side comment, it is important to note that unexpected functional values, if any, do not necessarily lead to what may be externally detected as an unusual use. Conversely, users may have attributed a functional value to the technology that is in line with its design rationale, and unexpected uses stem solely from a lack of exploration of the technology.

Let's now come back to the point that activity is always social and examine the implications when analyzing uses from a functional value perspective. As we saw in chapter 2, stating that activity is always social reflects the idea that (as demonstrated by the AT, structuration, and genre approaches) our individual—and, of course, collective—activities are influenced and/or shaped by social and cultural constructions, although we are often "ignorant consumers" thereof. We develop our conceptualizations, knowledge, or beliefs, but also our goals and needs, and thus our perception and meaning-making mechanisms, in sociocultural contexts. Far from being "neutral," technologies also reflect the historical and sociocultural context within which they have been imagined and designed and, as such, convey socially shaped worldviews and ways of doing.

The implication is not that our social-cultural context *determines* the functional values we attribute to technologies. The different use studies reporting unexpected or emergent functional values that I reviewed provide evidence for this. Rather, the implication is that, in some cases, fully understanding why we attribute (or not) certain functional values to certain technologies requires *considering if and how social forces play a role.*

A possible path is to address this issue by considering individuals as members of collectives and studying these collectives as such. The consideration of abstract collectives such as "young people" leads to only broad conclusions. In direct contrast, a detailed analysis of specific collectives (e.g., a given organization or a profession), their stability, and their evolution leads to an understanding of the social dimensions at play. When taking an AT perspective, a classic approach is to use Engeström's (1987, 2001, 2009) activity system model (AT "third generation"). Spinuzzi's (2003) genre tracing analytical approach is another means, which pertinently articulates the genre theory with an AT perspective on mediation and the different levels of activity in a given organization. The rationale for this approach is the empirical demonstration of how (1) workers' activities may be jointly mediated by ecologies of (rather than isolated) genres and (2) genre ecologies are both relatively stable and constantly evolving (e.g., via imports or hybridization) and include unofficial genres developed by the users in response to the evolution of the problems and/or the technology. This provides convincing evidence of the need to trace genres developmentally and, using genres as an integrated-scope unit of analysis, analyze how breakdowns or destabilizations at a given level (activities, actions, and/or operations) reverberate across the others.

The analytical path I propose—that is, focusing on the individual rather than the organization and studying if (or, rather, how) social structures and forces play a role in the attribution of functional values process—is clearly different. As already argued, I do not claim that focusing on collectives or organizations is meaningless or unproductive. On the contrary, I think that these studies are very helpful. However, the light that they shed on appropriation remains insufficient. Without rehashing the preceding discussions, my core arguments include how the technical ecosystem increasingly empowers individuals and the importance of the mediation (psychological) phenomena studied in chapters 2 and 3.

Kaptelinin and Nardi (2006) have nicely contrasted a similar difference in Leont'ev's vs. Engeström's perspectives on activity: (mostly) individual vs. collective form of activity, (mostly) individual vs. collective object owner, and motivation/need vs. production driving phenomena. Although the perspective adopted in this book does not downplay the interest of the latter, it is clearly closer to the former. Indeed, one may consider that a given group of friends or workers attributes a common functional value to an artifact or group of artifacts. However, if we analyze all the different aspects, forces, and processes reviewed (e.g., motivations, conceptualizations or meaning-making mechanisms), it is unlikely that a group can be considered as a coherent actor whose characteristics and activities explain what happens at the level of each different individual. As a consequence, considering the group level will inevitably lead to abandoning aspects that may be key.

In a nutshell, the key takeaways of this chapter include:

- the notions of *functional value* and *functional transparency*, and the importance of the former as an analytical tool;
- a precise definition of appropriation; and
- a general explanation of why and how appropriation develops.

In short, the proposed account is that the functional values that crystallize and lead to stable appropriations stem from how users interact with the tasks they consider. The different factors that play a role in these interactions therefore also play a role in what functional values users attribute (or not) to technologies, and shape why and how they do so. As we have seen, this includes the personal motivations or conceptualizations of users, the ecology of artifacts, the situated contexts where the interaction with the task takes place, and more general social and cultural forces.

What must be now clarified is the process that leads us to attribute functional values—that is, to use new means and/or new ways of doing things and extend our capacity to act. Chapter 4 first clarifies the psychological mechanisms underlying appropriation—that is, how it works—by studying users as humans who develop. Chapter 5 will then explore how, in addition to obvious forces such as work practices or social interactions (see chapter 2), unconscious and/or hidden functional needs stemming from our human nature may impact how and why we attribute (or not) functional values to technologies.

4 Users as Humans Who Develop

The fact that people develop is not a side effect of appropriation; it is the underlying mechanism. The extension of our repertoire of internal resources—that is, our psychological means (notions, knowledge, procedures) is a cognitive development. Appropriating a technology literally changes us. It is thus insufficient to state that users' interactions with artifacts may lead the users to evolve. Whether for emancipation concerns (enhancing individual control and agency) or design concerns (enhancing the likelihood that people can appropriate the technology), it is important to understand the mechanisms at play.

This chapter studies the psychological mechanisms underlying appropriation as addressed in chapter 3—that is, focusing on the development of the user's capacity to achieve tasks. I first present a general picture (section 4.1) and a psychological clarification of what the development of notions and patterned ways of doing/using correspond to (section 4.2). I then study the insights for understanding appropriation of addressing patterns as cognitive structures (section 4.3) and then consider the works studying the development of notions (section 4.4). Similarly to the other general processes I built on in the preceding chapters, the different developmental theories that are referred to in this chapter are addressed solely from the perspective of the appropriation matter of interest. For readers who do not wish (at least at the first reading) to consider these psychological details, the last section of the chapter provides a summary with the major inputs in terms of design. (However, as a core takeaway for design is the importance of user conceptualizations, I would strongly recommend reading section 4.4.)

4.1 Appropriation and Development: A General Picture

In their analysis of the use of apps such as messaging, accessing a social network, or sharing agendas, Bødker and Christiansen (2012) mention that, in their words, "[it's not only that users] draw the technology into their life world, they do in fact expand their life world and do things, they did not expect themselves. They see their capabilities in a new and expanded light and are able to do more than they expected" (p. 84). What this conclusion reflects is that a macro-perspective is generally needed to characterize and understand how the attribution of functional values to technologies and their inclusion in the ecology of artifacts modify users' capacity to achieve one or several tasks.

When taking an activity-centered approach, a classical means for conducting such macro-analyses is the expansive learning perspective. This approach, which builds on activity theory, posits that "an expansive transformation is accomplished when the object and motive of the activity are reconceptualized to embrace a radically wider horizon of possibilities than in the previous mode of the activity" (Engeström, 2001, p. 137). It requires considering the individual's activity systems and the group's or organization's activity systems within which this individual activity takes place. As we saw in chapter 2, these activities may be analyzed by examining the artifact mediation but, also, the community (the different actors) involved in the activity system(s), the rules and conventions regulating their actions, and how the activities are distributed among them. Issues and breakdowns are driving forces of change and development. Quoting Engeström (2001): "When an activity system adopts a new element from the outside (for example, a new technology or a new object), it often leads to an aggravated secondary contradiction where some old element (for example, the rules or the division of labor) collides with the new one. Such contradictions generate disturbances and conflicts, but also innovative attempts to change the activity" (p. 137).

As synthesized by Stevens and Pipek (2018), the expansive learning perspective leads to the consideration of two intertwined levels: "In respect of the individual level, [appropriation] refers to expansive learning and the formation of new routines, skills and competencies. Secondly, as a collective process it refers to the establishment, reproduction or transformation of action systems where new tools, rules, routines, and division of labor find their place within the activity system" (p. 144).

Developmental aspects must thus be studied at both individual and collective levels such as groups, companies, communities of practice, institutions, or societies. As an example of the latter, Malaurent and Karanasios (2020) analyzed how a subsidiary contextually adapted the enterprise system provided/imposed by the company headquarters. In the words of the authors, the local users "moved outside their own activities to self-organize into a supra-activity with the objectives of learning and creating workaround practices. . . . In practical terms, this led to the creation of user groups and a knowledge database in which users shared tacit knowledge about the enterprise system and ways to work around the misfits and frustrations identified" (p. 653). The company headquarters would later formalize some of the identified practices by recognizing the value of users' creation of new knowledge and new practices and integrating them within the system. As another example, Bardram (1997) showed how medical forms capitalize on years of experience, prompt for important information structures, and support hospital workers' activities. In Bardram's words, "The cognitive plans and their material counterpart are mere reflections of each other because they are both resources for, and products of, human activity" (p. 21). This actually leads these resources to be continuously (and collectively) modified and shaped.

Within this big picture, and in coherence with the perspective adopted thus far, the analyses developed in this chapter take the individual as an analytical entry point. The focus is on how individuals consider tasks and, in this context, attribute functional values to artifacts and develop specific uses. The role of social and collective processes is not downplayed but is considered from this perspective.

4.2   Development of Notions and Patterned Ways of Doing/Using

Notions, Knowledge, and Procedures

The psychological means that allow and drive our perceptions and actions (the internal resources of functional organs) include notions, knowledge, and procedures. These means are not innate. We develop them through experience and social interactions such as teaching, training, and, more generally, our different activities. This is why I refer to them as psychological constructions.

As a way to illustrate what is at play, let's consider how we create an account on e-commerce websites, browse photos on a social networking app, or format texts in word processors. Although we may now achieve these actions more or less transparently, they mobilize notions (e.g., "username," "password," "post," "like," or "text emphasis"), and require actions or series of actions (e.g., "swiping down an item" or "selecting a portion of text and then clicking on an icon such as italics or bold"). At some time we discovered and questioned or tested these notions and actions, then mobilized them in the course of our activities, and this led to our current personal understanding and practice of them. In other words, we developed our own notion of "account," "profile," or "text emphasis," which is why it may differ from that of other individuals and may be more or less congruent with that implemented in the application (i.e., what the designer would consider as the "real" or "correct" notion). Similarly, we developed the more or less automated series of actions allowing transparent use: the way we address the task and use the technology progressively stabilized and then crystallized.

This elaboration of new concepts and knowledge does not occur in a vacuum. It is based on, or at the very least influenced by, the concepts and knowledge we had previously developed (in Piagetan terms, what is at play includes assimilation and accommodation mechanisms). As a basic example, most social networks feature notions such as "posts," "followers," and "like" notions. Once one has developed the concepts and procedures for an app, using a new app based on the same or close notions may be straightforward. Now let's consider a more complex example. When using a word processor such as Word or LibreOffice Writer, knowledge such as "selecting a portion of text and clicking on the B button makes the text bold" mobilizes notions such as "portion of text" or "bold." This is arguably basic: The notions are simple; what they refer to is straightforwardly related to text management and can be seen on the screen. However, consider knowledge such as "characters have editable properties (including bold, italics, font, or size)," which mobilizes the abstract notion of "property." The understanding of the notion of "property for characters" that users develop is likely to be related to their previous understandings of the general notion of "property" (e.g., extension of the initial notion) and to whether or not they are already aware that "spacing" or "indents" may be regarded as "paragraph properties." Similarly, the notion of "style" is a complex and abstract notion, the development of which is related to that of other notions such as "properties," "character," "paragraph," or "spacing" (the latter may support the development of the former, or vice versa).

It must therefore be kept in mind that, whatever the way we encounter them (e.g., a tutorial, help from a peer, and/or exploration), the notions that drive our perception and meaning-making mechanisms, and, in turn, drive our ways of doing things and of using technologies, are an output of complex and idiosyncratic psychological developments. We do not "create" new notions from nothing, and they are not "transferred" into our heads by teachers, peers, or designers. We develop (via, in particular, our activities) and, in this context, shape, adapt, extend, and/or specialize notions and knowledge.

As a striking example, see the videos of babies attempting to turn the pages of paper ("real") books by pressing/swiping them: They perceive and interpret the setting according to the notions and processes they developed from their uses of smartphones or tablets and act accordingly. Although the *digital native* label has little scientific meaning, this may be one: Some humans now perceive and conceptualize some of our basic tangible objects in a way that is impacted and/or shaped by the conceptualizations developed via their digital experience.

Patterns and Schemes

As we saw in chapters 2 and 3, all works considering appropriation acknowledge the role of patterned ways of doing things or using technologies. Acknowledging the role of such patterns is not a claim that our actions are driven by predefined fixed and rigid structures that we mechanically apply. The work conducted by Suchman (1987), and many others, highlights that action is always situated. Nevertheless, as argued by Orlikowski (2000), "While a practice lens recognizes that technology use is always situated and emergent, it does not imply that such use is completely unique. On the contrary, because regular use of the same technology tends to be recurrent, people tend to enact the same or similar technologies-in-practice over time" (p. 421).

In cognitive sciences, the research tradition addressing the patterned nature of some aspects of human behavior includes very different perspectives (and is fairly controversial). For instance, the existence of large and stable cognitive structures posited by the general script theory (Schank & Abelson, 1977) has been both historically influential and very criticized. Other approaches, including that of Schank (1999), adopt a more flexible and dynamic perspective within which finer-grained components are contextually combined according to the situation and the individual's goals. Some of these articulate scripting and sociocultural perspectives (see for example Fischer et al., 2013).

With respect to appropriation, how different works have proposed to address such patterns as *mental schemes* (schemes for short) is of specific interest. The notion of scheme, which seems to originate from Kant, is central to the work of the psychological development theorist Piaget (1952). Piaget's notion of scheme is usually referred to as a cohesive, repeatable action sequence possessing component actions that are tightly interconnected and governed by a core meaning. Building on Piaget's perspective, Vergnaud (1998) defines a scheme as an invariant organization of behavior for a certain class of situations. Within this perspective, the observable part of schemes is the more or less stable sequence of actions by people facing instances of the class of situations.

The explanation proposed here is thus that the patterns that may be observed when analyzing how people conduct their tasks or use their apps reflect—and are an output of—(1) the classes of situations they have developed, which drive their perception and interpretation of settings, and (2) the associated schemes (the associated ways of doing/using). The empirical works conducted in this light in the context of teaching academic (Vergnaud, 2009) and professional skills (Rabardel & Pastré, 2005; Pastré et al., 2006) show that the concepts that people mobilize and what people consider to be true both play a key role in the recognition of situations. These concepts and beliefs may, however, be highly idiosyncratic. They may differ from what is (or is considered as) "correct" and, more generally, from expectations.

What is specifically interesting about this perspective is its congruence with the consideration of users as pragmatic actors (actors who do) rather than epistemic actors (actors who know) that is developed in this book. These works do not claim that, when we consider a task or use a new technology, we first explicitly consider our repertoire of known and precisely defined classes of situations or technology interfaces, then analytically examine which one best matches the setting we face, and finally decide how we should act. As shown by the empirical studies conducted by these authors, classes of situations are defined with varying levels of precision and actors may not be able to explicitly describe the situations they implicitly refer to, their definitions, or their scopes. Nevertheless, for some of their classes of situation, the schemes developed allow actors to recognize and address tasks in a relatively straightforward way. (For others, they have to explicitly think and explore possibilities.) For instance, for users having developed a "sending emails" class of situation, using their usual application but, also, a new application, may be transparent or relatively straightforward. A new situation—for example, sending messages in a collaborative platform or posting in a social network—may seem completely new or, on the contrary, be (erroneously) recognized and interpreted in the light of the existing class of situation, which may lead to issues (as the situation and technologies are indeed different). This may lead to the development of a new class of situation and/ or a refinement of the original one that avoids some erroneous recognitions.

The tentative models proposed by theorists attempting to make sense of how people develop skills illustrate what may be the "ingredients" of schemes. The model proposed by the theory of conceptual fields (Vergnaud, 2009) resonates well with many of the analyses proposed in this book. Vergnaud defines a scheme as a complex structure comprising intentional, generative, epistemic, and computational aspects. Intentional aspects include goals, subgoals, and anticipations. Generative aspects involve "rules to generate activity, namely the sequences of actions, information gathering, and controls." Epistemic aspects (i.e., the concepts that are mobilized and what is considered to be true by the actor) allow to "pick up and select the relevant information." Finally, computational aspects provide possibilities of inference "which are used to generate goals, subgoals and rules, or properties and relationships that are not observable, in particular in new situations." Schemes thus contain both "ready-made rules, tricks and procedures that have been shaped by already mastered situations" and means to adapt to new situations (p. 88).

Implications

The studies reported in this section lead to two important points. First, taking a cognitive perspective, our uses of technologies stem from psychological constructions (notions, knowledge, and processes) that we develop, and, in turn, these uses trigger or participate in this developmental process. Second, these psychological constructions are, together with the resulting behaviors, stable-for-now structures that can and do evolve. In other words, behaviors are simultaneously patterned, adaptable (actors/users can address variations in tasks and conditions), and evolutive.

As raised by Vergnaud (1998), what is invariant is not the action but the organization of action. For instance, actions for sending an email, posting in a social network, using a collaborative platform, or editing texts in word processors may vary from one app or device to another, but the organization of these actions and the set of underlying notions (e.g., "message," "sender," "like," "channel," "drive," or "style") is relatively stable. Moreover, invariance is to be understood as invariance at a given moment or, in other words, at a given stage of development. People develop invariant ways of doing and mobilize them as such. Nevertheless, these ways of doing may and do evolve, breakdowns being a typical context for adaptations.

Let's now study how these points (1) provide a better understanding of the perception of action possibilities, of user/technology coevolution, and of importing ways of doing; and (2) highlight the need for caution when considering the notions underlying the use of technologies.

4.3   Perception of Action Possibilities, Coevolution, and Import of Ways of Doing

Addressing patterned ways of achieving tasks and using technologies as cognitive structures provides a psychological account for three important aspects: perception of action possibilities, coevolution, and import of ways of doing.

Perception of Action Possibilities and Attribution of Functional Values
I argued in chapter 3 that the attribution of functional values to technologies is driven by how people establish relations between their activity—that is, the tasks they consider, and how they consider them, and the action possibilities of technologies (or, in other words, the effecter affordances) that they perceive.

This key idea can be found in different works originating from different backgrounds and may thus be considered as a consensus. As we saw, the activity-based approaches of the notion of affordance argue that the users' perception of what artifacts may offer is shaped by their objectives and activities. When considering how affordance operates, Turner (2005) mentions that "in their own very different ways, both Ilyenkov and Heidegger have argued that we understand the world in terms of use" (p. 798). The structural approach to affordances acknowledges that what the technology does and symbolizes will emerge over time as a direct result of how users adapt and restructure their working practices (Vyas et al., 2017). Similarly, Orlikowski (2000) highlights that a technology-in-practice serves essentially as a behavioral and interpretive template for people's situated use of the technology. In the same vein, the sociology of organizations perspective developed by Barley (1988) argues that "workers attempt to assimilate new technologies under previous patterns of practice and interpretation" (p. 50).

Although addressing such processes in terms of rigid cognitive structures is obviously misleading, the scheme perspective provides, as we have seen in section 4.2, a theoretical account, draws attention to the interplay of intentional, generative, epistemic, and computational aspects and suggests that we should focus on classes of situation rather than on precise settings and their contingencies.

Interestingly, this account also resonates with the genre perspective. The congruence is illustrated well by the previously mentioned quotes of Bazerman (1994): "Individuals perceive homologies in circumstances that encourage them to see these as occasions for similar kinds of utterances," and "Genres, in-so-far as they identify a repertoire of actions that may be taken in a set of circumstances, identify the possible intentions one may have" (p. 69); Spinuzzi (2003): "[Genres are] learned habits or responses on which a worker unconsciously draws" (p. 144); or Russell (2009): "[Genres are] a typified, tool-mediated response to conditions recognized by participants as recurring" (p. 43). See also Coe (1994): "A genre may be understood as an incipient action, i.e. a potential action waiting for an activating situation. When we see past the uniqueness of a particular situation and recognize it as familiar, we activate (at least provisionally) a structure we have previously decided is generally appropriate to that type of situation" (p. 155). As mentioned by Russell (2009), genre (as social action) can be analyzed as both explicit strategies (for instance, Spinuzzi [2003] has shown how workers consciously selected, rejected, or abandoned genres and their associated technologies in the course of their work) and as unconscious operations or activity systems.

The genre approach specifically adds the highlight of a dialogical dimension. As discussed by Bazerman (2013) in his analysis of language development, "Prior experiences of language become formative elements in the development of individual thought—not by direct importation of a language symbol or ideological system, but because the child first interacts with the language in the course of activity and then redeploys that language as part of self-regulation in tasks including his or her own interactions with others"

(p. 32). If we move to genres as such, Miller (1994) notes that "social actors create recurrence in their actions by reproducing the structural aspects of institutions, by using available structures as the medium of their action and thereby producing those structures again as virtual outcomes, available for further memory, interpretation, and use" (p. 71).

What surfaces here is that genres must be learned. This learning process occurs via interaction with others and/or technologies reifying some aspects of the genre (using AT terminology, Russell [2009] notes that genres may be "potentially passing from the level of action to operation and back" [p. 44]). However, the practice of the genre and how it fulfills individual needs also leads users to create personal meanings, which influence perceptions and, in turn, action. In other words, while interacting with the tasks we consider and the technologies through which we address them, we *develop* our own understanding of the genre (e.g., its characteristics, the setting it applies to or the conventionalized way of acting). This understanding is saved from solipsism by our constant interactions with others but may nevertheless lead to idiosyncratic perceptions. As an output, we may fail to perceive the genre at play and thus not benefit from the associated knowledge or, on the contrary, interpret new settings/designs in an innovative way.

Coevolution

The second interest of addressing patterned behaviors in terms of cognitive structures is to suggest considering the quadruplet (task, way of doing, artifact, way of using)—that is, the task that is considered by the individual (or the set of tasks when the considered one interplays with others); the cognitive structures shaping the patterned way(s) of addressing these tasks (the associated schemes); the associated artifact(s); and, finally, the cognitive structures shaping the patterned way(s) of using these artifacts (i.e., the use schemes). The artifact-use scheme couple is to be regarded in the light of the task-scheme couple. The situation, as recognized by the user and his/her associated schemes, forms the context within which the technology is used as a resource.

From an analytical perspective, considering this quadruplet offers an instrumental grasp of the general descriptive stance according to which people both adapt to technology and adapt technology to themselves. What is adapted may relate to the task (users' conceptualization of the task and the domain, users' way of doing); the artifact (users' conceptualization of the technology, users' way of using, users' technical adaptations); and/or the link task-artifact (users' attributed functional values). All three aspects include user cognitive development.

This understanding provides a theoretical account for the point according to which users' adaptations of the artifact are only a contingent aspect of appropriation. What is intrinsic to appropriation is the emergence of this quadruplet and the according user evolution. Technical adaptations, if any, occur in relation to the development of the ways of doing/using and are driven by this development.

This understanding also provides a theoretical account and an instrumental frame of mind for how appropriation may develop in relation to both new and previously used artifacts, and how these processes interplay. The appropriation of new artifacts corresponds to offered or imposed new technologies that are given functional values in relation to preexisting or emerging ways of doing. In the case of previously used artifacts, the evolution of personal or collective ways of doing causes the functional values of the existing artifacts and/or the ways of using them to evolve. These two forms of evolution may interrelate in different ways (Tchounikine, 2017): Old artifacts may be given new functional values or stripped of previous functional values independently of and/or in connection with the (new) functional values of other (old or new) artifacts, and, as a consequence, users may develop new and/or enhanced resources from these artifacts. This may stem from the evolution of an attributed functional value or from the way the user uses the artifact to achieve his/her goals based on this value. Rather than individual resources, what evolves may also be the way users relate these resources to each other— that is, an evolution as a system.

Import of Ways of Doing

Finally, another interest of considering the cognitive structures underlying users' patterned ways of doing and the task-way of doing couple is to draw attention to the stability of this couple. This stability can help to make sense of the import of usual procedures in new settings.

As highlighted by many authors since the very first works studying appropriation, the use patterns that people develop for one technology influence the way they use new technologies, with many users attempting to make the new operate like the old (Mackay, 1990). We saw several examples in the preceding chapters, including cases where users customize their apps as an output of their practices and, when moving on to new means, customize the new apps according to how they had customized the previous ones.

This point is well illustrated by the learning scenario editor example mentioned in chapter 3. Unlike traditional instructional languages that impose a given conceptual perspective, it allows teachers to adopt the structure that best fits the considered scenario (Sobreira & Tchounikine, 2012, 2015). For instance, teachers may contextually adopt an activity-group-resources representation (i.e., represent a scenario as a set of activities to be performed by groups sharing some resources) or a participant-role-activity-resource representation (i.e., represent a scenario as a set of participants playing roles associated with activities and resources). However, most of the tested teachers directly organized the conceptual notions in a certain order—that is, independently from the particular scenario they modeled, with many of them commenting that "I always [model scenarios] this way" and, in some cases, using complex workarounds to stick to their usual way of doing (Sobreira & Tchounikine, 2012).

This example illustrates how the users' ways of using the system is influenced by the previously constructed cognitive structures. As they are offered some flexibility, users tend to stick to their idiosyncratic perspective—that is, they attempt to do things in the way they have always done them and use the adaptation means to align the artifact with their practice. As we have already highlighted, the goal of the teachers is to address a pedagogical task, not to use the system. The driving forces are the procedures they developed via their pedagogical experiences and idiosyncratic practices, not the artifact features.

This aspect has been studied in detail in the genre approach. For instance, Yates and Orlikowski (2002) note that "in moving their communication to a new medium, members of a team or community typically import existing genres and genre systems, improvise around them, and gradually learn to take advantage of new opportunities afforded by the medium" (p. 32). Any interaction of users with an application, including the first one ever used, is impacted and possibly shaped by how these users habitually enact the genre (see the discussion on not-selfies in section 2.4). This gives specific interest to the genre tracing approach proposed by Spinuzzi (2003).

The scheme approach offers a straightforward psychological understanding: Users' existing schemes (way of addressing a task, ways of using a technology) influence and/or shape how they behave in a new setting. Initial and possibly long-lasting uses of a new application should thus be analyzed in the light of a possible import of existing schemes for both the task achievement and the application use. Typically, in the case of the learning scenario editor, most teachers stick to their existing task-related scheme and import their usual way of representing learning scenarios to the new technical setting they are offered, including when this does not work well. This process also likely applies to, for instance, the use of collaborative platforms such as Slack. When workers who are used to open team communication want the discussions to be visible via channels, those who are more accustomed to one-to-one communication and hierarchical organization tend to continue sending personal messages. In fact, all of these workers have developed their own collaborative classes of situations, which have been shaped by an interplay of forces including individuals' characteristics, local work practices, the used technologies, and/or more general occupational or social forces. Rather than the intrinsic properties of the newly offered communication platform and the social pressure for exploiting them, the way they engage in communication is likely to be shaped by the schemes associated to the classes of situations that frame their meaning-making mechanisms and consequent behaviors. This may indeed still be the case once they have perceived that some aspects of the setting differ from preceding experiences and remain until new psychological developments occur.

As a side comment it may be noticed that in above examples, what makes the import of ways of doing salient is that it does not work well. When the import does not raise any tension, or the tension is not identified, it is just not noticed. As already mentioned, odd appropriations or breakdowns highlight phenomena that are much more frequent than these cases suggest.

One-off appropriations may also be seen in terms of procedure imports. As we saw, functional transparency shapes the contextual search for new means. What is at play here are the schemes associated to the usual ways of doing and the habitually used artifacts.

The conceptual part of schemes plays a core role in this import of procedures. They pave the way or, on the contrary, create obstacles to, the adoption and appropriation of new technologies (or, as already mentioned, the perception of affordances). The recognition of the setting, of its semantics (more precisely, the mobilization of the meaning-making mechanisms shaping the semantic interpretation), of what may/should be done, and of the effects are shaped by the personal conceptualizations developed through previous practices.

A Possible Articulation: The Instrumental Genesis Theory

A specifically interesting articulation of several of the points made in this section has been proposed by Rabardel (2001, 2003). This author built on the ergonomics research tradition and the scheme perspective to propose an original understanding of the development of instruments, namely the instrumental genesis theory. As already mentioned in chapter 1, this perspective is a very interesting complement to the activity theory perspective on mediation, which can be better explained now that we have studied the notion of scheme.

Rabardel proposes to consider the couple (artifact + way of using) in the light of the notions of instrument, instrumentation, and instrumentalization, for which he proposes specific definitions. An instrument is defined as a mixed functional entity combining a technical dimension (the artifact) and a psychological dimension (the associated schemes). The development of instruments (namely the instrumental genesis) is a dual process comprising instrumentation—that is, the user's adaptation to the artifact constraints—and instrumentalization—that is, the attribution of functions to the artifact and the technical transformation of the artifacts by the user, if any. Instrumentation thus relates to the genesis of the human side of the instrument by developing, adapting, or reorganizing schemes. Instrumentalization is based on the initial attributes and properties of the artifact but extends the initial design by new local or lasting properties in accordance with the current action and situation (Béguin, 2007; Folcher, 2003). Here again, this perspective highlights how users' activities are both productive (users accomplish tasks) and constructive (they elaborate resources).

Instrumentation and instrumentalization processes are closely related to the professional dimension of the activity and how individual processes join collective practices: Instruments, as addressed here, are both individual and collective constructions. And the import of procedures (e.g., the case of the teachers organizing the conceptual notions of the editor in a certain order "to model scenarios as they always do") may be interpreted as the impact of past instrumental geneses.

Implications

In this section, we saw how patterned ways of doing shape perception and the potential import of procedures. I have focused on the psychological dimension of this phenomena, but, as we have seen, this perspective underlies other approaches such as the genre theory. As highlighted by Bazerman (2013) in his analysis of the pragmatic tradition within US social science, commonalities include that "human belief, knowledge, and perception were always interpretive" and that "the interpretations come not only from the social and historical position of the person, but from their engagement in projects to satisfy their needs, desires, and value-laden senses of fulfillment" (p. 88).

From an analytical and design point of view, the implications are twofold. First, any analysis of how people use technologies must pay specific attention to the preexisting constructions. The notion of scheme offers one possible psychological theoretical account and model. Second, as users' conceptualizations play a core role in these different aspects, they must be considered and analyzed as such. Conceptualizations are central to how users make sense of the setting and, according to Vergnaud, key to the generalizable part of the activity—that is, the capacity to adapt activity to the variety of situations.

Before we study in greater depth the role of conceptualizations, it may be worth taking a step aside and preventing a possible misleading positivist interpretation of the analyses developed above. As mentioned in chapter 1, the focus on activity—that is, what is done—reveals the need to consider users as pragmatic rather than epistemic actors. What is considered here is thus conceptualization within action, and not specifically what users know. The occupational didactics tradition shares this perspective and analyzes the subject who says "I can" before "I know" (Pastré et al., 2006). This is also congruent with Engeström's perspective to transformative action. As highlighted by Clot (2009), these approaches share a clear-cut epistemological position that also underlies this book: "[The] positivist catechism adheres to the principle From science comes foresight, from foresight comes action. In other words, this principle emphasizes knowing in order to foresee before acting. . . . Yet there is indeed an alternative to positivism that leads not to a weakening of scientific activity but rather to a bigger demand for it. Its principle could be the following: acting, without being able to foresee everything, in order to know" (p. 287).

4.4 Being Attentive to the Conceptualizations Involved

A conceptualization is a differential system of notions. As we saw in sections 4.1 and 4.2., our conceptualizations of domains, settings, devices, or features and their interrelations form the substratum of our perceptions, meaning-making mechanisms, and actions.

Any application or device is based on a conceptualization and involves more or less complex or usual notions. For instance, the use of email requires people to identify, dissociate, and understand notions such as "sender," "receiver," and "message," which are arguably basic notions. In direct contrast, what makes the use of image editors such as Photoshop or Gimp much more difficult is the mastering of the dozens of complex notions underlying their features. The fact a notion is "basic" is however of course relative. What a "channel" or a "follower" corresponds to, or the very notion of "social network," may appear basic for some individuals and complex, if not cryptic, for some others.

It is thus very clear that how we develop notions plays a role in the adoption and appropriation of technologies. In particular, non-adoptions, non-appropriations, or odd-appropriations may stem from, or be reinforced by, a mismatch between the user's developed conceptualizations of the domain, the task at hand, and the conceptualizations underlying the used application. *Mismatch* is used here rather than *misunderstanding* to avoid a value-based judgment.

The role of user conceptualizations has also been demonstrated in different studies that do not take a cognitive perspective, as I do in this chapter. For instance, the study of repurposive appropriation of digital cameras conducted by Salovaara et al. (2011) revealed how what these authors call "technology cognizance" ("having a comprehensive and correct mental model of a camera") makes it easier to notice opportunities in novel situations. This idea may also be found in approaches featuring the role of structural properties. For instance, Orlikowski's (1992) analysis of the adoption and appropriation of the Lotus groupware revealed the role of people's mental models (their model of the technology, their model of the job tasks) in the way they used the system.

While studying how we develop and mobilize notions as such would take us off course, let's quickly review how some of the points made in psychology and learning sciences help understand the appropriation (or non-appropriation) of technologies.

Preexisting Conceptualizations Play a Key Role

A first important point is that activity is never free of conceptualizations and principles. The primacy of action I put forward and the consideration of users as pragmatic actors (actors who do) rather than epistemic actors (actors who know) is not to be understood as a statement that actors who do do not have any conceptualizations. Works conducted in the learning sciences field have shown that students' incorrect ways of solving problems may seem erratic but are often understandable in terms of incorrect conceptualizations.

This point deserves specific attention insofar that new technologies are generally introduced for practices that are already mediated by other digital or nondigital means, and thus for which the (future) users have preexisting conceptualizations. The introduction of any technology, including innovative ones that apparently create brand new possibilities, must be analyzed in this light.

As an illustration, let's consider social networks and collaborative applications. It can be argued that these applications create a new communication ecosystem. They not only facilitate the sharing of texts and images; they also introduce an interaction that is somewhat graph-based rather than one-to-one or one-to-many and put forward new notions such as "contacts," "followers," "likes," or "stories." Is it possible that these practices developed ex nihilo? This would be quite a positivist line of thinking. The actual central issue with social networking apps is socialization and its general functioning. Humans perceive the new in the light of the old, and a technology cannot be perceived in a way which is independent from previous activities, previously used technologies, and the notions that underlie both of the aforementioned. We will come back to this point in section 7.5 when studying the interests of favoring productive analogies.

Humans Can and Do Deal with Implicit and/or Imprecise Notions

A second important point is that individual's conceptualizations and beliefs are often informal and/or implicit; what is salient is their enaction only (Vergnaud, 2009). This point sheds some light on issues such as "The user is unable to explain what frames the way he/she acts," as irritated designers may sometimes complain.

The fact that we (humans) often act within informal conceptualizations is not an issue as such. However, it may be difficult to reconcile with the way technologies impose predefined structures/conceptualizations. As an illustration, most if not all email clients offer more or less identical ready-to-use features to define and manage tasks, events, and calendars. For instance, the interface for defining a task in the Thunderbird (2024) email system features slots such as "Title," "Location," "Category," "Start," "Due Date," "Status," "Repeat," "Reminder," "Description," or "Attachments." There is little doubt that most users can understand the notions underlying this task-management feature and learn how to use it. However, empirical studies such as Pucihar et al. (2016) reveal how, while notions as tasks or projects may be clearly defined and structured in corporate or institutional contexts, it is usually very different on a personal level. Many users do not conceptualize their activity in terms of tasks or use this notion in a very flexible way.

What may become an issue here is thus not a matter of conceptual understanding but a matter of *conceptual flexibility* (we will see how to deal with such issues in section 7.2). Here and now, when one wants to focus on one's activities and not on the application, the informal, situated, and adaptive task notion that we use in practice may not correspond to that of the application, or may only appear to do so, and may be difficult to project onto the technical device.

How humans can and do deal with implicit and/or imprecise notions thus also sheds some light on issues such as "The user is unable to understand the notions proposed by the system," as here again irritated designers may sometimes complain. It is important to dissociate the understanding of a notion and its use, in action, as a psychological tool.

Users' Zone of Proximal Development

Finally, a third important point is that users cannot be expected to act/develop outside their *zone of proximal development* (ZPD). The ZPD is a concept that has been used as a starting point for a variety of studies in developmental and educational research. It is defined as "the distance between the actual developmental level as determined by independent problem solving and the level of potential development as determined through problem solving under adult guidance, or in collaboration with more capable peers" (Vygotsky, 1978, p. 86). This is often illustrated by three concentric circles featuring (from the inside to the outside) (1) what the individual can do alone—that is, the current state, (2) what the individual can do if aided by a tutor or a more capable peer, and finally (3) what is beyond the reach of the individual at present.

As an example of how the ZPD notion may help understand the evolution of practices, let's consider the case of users using emails to remember things they have to do. Potential developments include (1) more efficient technical implementations of present resources (e.g., a more systematic use of filters, filing, and tags to reduce the inbox volume and allow to-do emails to be more visible) and (2) skipping to a task management feature, which may require a conceptual development. Taking a ZPD perspective helps us to understand that, counterintuitively, the latter may thus prove to be more demanding than the complex technical workarounds requested to keep the present conceptual perspective reasonably efficient.

The implication is that designing for appropriation requires paying attention to the conditions of user development, namely the key role of conceptualizations, what is possible and what is too far, and the collective dimension. The way the ZPD notion features the role of more capable peers echoes the studies pinpointing the role of other users in appropriation and adaptation of technologies. Here again, we will see how to deal with such issues in chapter 7.

4.5   Synthesis and Implications

Acknowledging that appropriation is fundamentally a developmental process allows the fine-tuning of some of the claims made in chapter 3. In particular, addressing ways of doing and ways of using as psychological constructions, for instance as schemes, makes it possible to address functional and use transparency as the outputs of the different schemes that the user developed and their fit with the considered setting and applications. When it is coherent with the schemes (i.e., with the ways of doing) that the user has developed for a given task, a new application may immediately become functionally transparent. Its ease of use and, possibly, immediate or forthcoming use transparency, is another question, which partly relates to whether the use schemes associated to the previous technology remain suitable for the new application and, of course, the design and general usability of this new application. This may lead to adaptations—that is, users align the artifact with their use schemes as far as they can, and/or schemes adaptations (see Rabardel's model). Use difficulties or breakdowns, however, may also impact the functional task–application link. As put forward by constructivist perspectives, tensions create necessities and thus opportunities for developments. Unfortunately, they may also cause damage.

This understanding, which resonates with the notion of functional organs introduced in chapter 2 ("functionally integrated, goal-oriented configurations of internal and external resources"), provides a theoretical basis for the claim, first made in chapter 3, that the analysis-driving axis of appropriation must be that users interact with tasks and, in this context, create resources.

This understanding also highlights the risk of overinterpretations and, in particular, of labeling any successful use of a new technology as appropriation. This point already surfaced when we considered one-off appropriations. Developmental processes are generally long-term processes. What is occurring when we use a large screw to open a bottle of wine is a contingent episode based on an existing, imported, and more or less adapted way of doing. This is very different from the long-term process of developing ways of remembering things we have to do via the email, although one-off episodes may be the origin of these processes. As another example, it takes a matter of minutes to see and understand what the task notion corresponds to in a PIM application, while integrating and using it in our basic everyday practice may take months. Although technology designers or promoters sometimes tend to interpret any successful use as appropriation, it is not always the case.

Finally, this understanding provides a theoretical basis for the claim, first made in chapter 1, that the degree to which the designed features of a system are consistent with the users' actual activities cannot be considered statically. Development is an ontological characteristic of humans, and, as we saw, activities are both productive (achievement of tasks) and constructive (elaboration of resources). The way we conduct activities and use technologies stems from past developments (notions, more or less patterned ways of addressing tasks and using technologies) but also nurtures the continuing development thereof. A given set of technical features may thus be in line with our activities at one time but only partially support them and/ or become obstacles at a later date.

In addition to providing a theoretical basis for the abovementioned claims, the analyses developed in this chapter have pragmatic implications.

- Users must be considered as pragmatic rather than epistemic actors, and what is at play is their conceptualizations within actions rather than their knowledge.
- It is important to consider users' preexisting (i.e., previously developed) notions and patterned ways of doing/using: They shape their perception of action possibilities and their import of procedures and ways of doing.

- It is important to dissociate the understanding of a notion and its use, in action, as a psychological tool.
- Users cannot be expected to use means that require them to master notions or processes that are not in their zone of proximal development.
- When humans can and do deal with implicit and/or imprecise notions, the way designs define particular definitions/implementations of the application domain notions may be an issue.
- Even in the hypothetical case of considering a single user with a holistic and precise requirements analysis, a smart design, and positive initial use tests, the constructive dimension of appropriation may lead to user-artifact tensions.

5   Users as Human Beings

The perspective adopted in this book is to consider users as actors who act in and on the world and, in this context, use and appropriate technologies. However, before being users, we are human beings. We saw in chapter 2 how obvious social forces such as work practices or collectives (e.g., groups of friends) may play a role in appropriation processes. What I want to present in this chapter is an idea that has been underexplored in the works considering technologies appropriation, if not completely ignored: Psychological forces and needs sometimes lead people to engage in specific high-level activities that generate, or interplay with, digitally mediated lower-level activities. As we will see, users' motives and activities may thus be multiple, relate to different levels, and be highly unexpected. Acknowledging this complexity is core. An open-minded holistic perspective is thus necessary, and, as I do not claim completeness or perfection, the forces I will present in this chapter may also suggest and pave the way for future analyses.

The analytical axis I adopt in this chapter sticks to the general activity-centered perspective adopted in the book and explores how our human nature leads us to consider (often unconscious) specific goals, and thus to engage in (often unconscious) specific activities. This analytical approach is thus different from studying how our personality traits and psychological profiles impact our digital activities, which is a well-studied research topic. A large set of empirical studies have studied this impact, and their findings are also useful to the deciphering of some technology uses. As examples of findings, the lack of control resulting from the email "overload issue" (Dabbish & Kraut, 2006; Grevet et al., 2014) and, in particular, the risk of missing an important email, often generates high levels of stress and anxiety (Hanrahan et al., 2016; Reinke & Chamorro-Premuzic, 2014) and shapes or influences some practices. Similarly, FOMO (fear of missing out)—that is, the apprehension of missing a rewarding experience—is both an output and a driver of the addictive use of smartphone and communication apps, and low levels of life satisfaction and self-esteem seem to be both a cause and a consequence of Facebook addiction (Błachnio et al., 2016). As a final example, it has been shown that the big five personality traits (extraversion, agreeableness, openness, conscientiousness, and neuroticism) play a role in how people perceive and thus use self-presentation technologies such as LinkedIn, Facebook, Twitter, Instagram, Tumblr, or Snapchat (de Vito et al., 2017).

I will first present empirical studies providing evidence that we sometimes engage in activities driven by unconscious and/or hidden functional needs stemming from our human nature, and show how the works addressing human motivation facilitate the understanding of such activities (section 5.1). I will then take a top-down theoretical perspective driven by the following line of thinking: Given that such cases do exist, the general theories studying our human nature may facilitate the identification and deciphering of such occurrences. To explore this idea, I will study if and how existential needs (section 5.2), psychological attachments (section 5.3), and engagement (section 5.4) may help to explain how we use some technologies, and present a selection of examples that provide confirmation that this is indeed the case.

Although the work conducted here is partly speculative, it clearly suggests that a full understanding of how we use and appropriate technologies does require the consideration of these general forces. The conclusion is that empirical works confirming these insights and allowing a better understanding of this phenomenon should be conducted in the future, for the theories I review here, and more generally. This will be discussed in section 5.5.

5.1   From Psychological Factors to Psychological Generators

Examples
Let's start by illustrating how high-level general psychological needs sometimes generate specific activities.

Avoiding thoughts   Wilson et al. (2014) conducted a series of empirical studies where participants were asked to spend six to fifteen minutes entertaining themselves with their thoughts only, with no access to any source of distraction. The study found that some individuals find it unpleasant to spend time with nothing other to do but think. Actually, when offered a device that produced an electric shock by pressing a button, many of the individuals preferred to press the button rather than being left alone with their thoughts.

This unusual study strikingly illustrates how a psychological need may generate motives and, in turn, a specific activity and the consequent mobilization of a mediator. In some sense, these participants attributed a contextual "avoid facing my thoughts" functional value to the electric shock device.

Although this case is obviously a one-off episode, it may reasonably be hypothesized that, for those who don't want to be left alone with their thoughts, this high-level "avoid facing my thoughts" functional value is contextually attributed to other means. Along with other reasons such as dealing with anxiety (see the introduction of this chapter), this may be part of the explanation for how people interact with their smartphones: In their use study, Heitmayer and Lahlou (2021) found that most of the interactions with the smartphones were initiated by users rather than by a notification from the system. It may also help to make sense of how, after having noticed the absence of a new message to deal with, some people will spend the few remaining minutes before arriving at their subway station playing video games designed for six-year-old children.

Adapting applications   All the works studying technologies appropriation agree that the way people act through computers to address their jobs or tasks often leads them to adapt applications to their needs and uses. Chapter 6 studies this point in detail.

However, the most frequent type of technical adaptation is probably changing the original "skin" or "theme" of devices, operating systems, and/ or applications—that is, their graphical appearance. The interest (I would say, the perceived usefulness) of such features is well illustrated by the thousands of different skins available for some popular applications and their high download figures.

While changing the appearance of applications, the background image of the computer desktop, or the ringtone of the smartphone is undoubtedly a way to make the technology our own, the reason people take time for such activities cannot be explained by the alignment of the technology with usual ways of conducting tasks or the pursuit of performance. Such changes do not modify the system features in any way, and the rationale should be sought in other types of motivations.

Among many other studies, Blom and Monk's (2003) investigation of why users personalize their computers and mobile phones reveals that these motivations include aesthetics, reflection on personal or group identity, or the eliciting of emotional responses. As another example, during the Covid pandemic, people used videoconferencing apps such as Zoom to conduct their work practice and, in this context, expressed aspects of their identity or emotions via contextually chosen video backgrounds.

As stated by Tractinsky (2004), aesthetics are a basic human need and, for many users, are the most important aspect of their interaction with technologies. As a consequence, aesthetics may supersede any considerations of functionality or alignment with practices when selecting or customizing devices and applications. For instance, it was found that some gamers are more likely to spend money to customize the appearance of their avatar in the game than to improve said avatar's abilities (Rogers & Dunlow, 2019). In other words, the appearance is more important than gaining a functional advantage.

Making Sense of Such Cases

Why people engage in activities such as avoiding thoughts or changing the appearance of their applications or devices may be considered in the light of the general theories addressing humans' needs and behaviors. This is well illustrated by Oulasvirta and Blom's (2008) use of the self-determination theory (SDT; Deci & Ryan, 2000). SDT is a macro theory of human motivation and personality arguing that humans face three psychological needs: competence, autonomy, and relatedness. Competence is related to our capacity to effectively act on the world, and to our need to have a personal impact on the environment, self, and others. Autonomy is related to our desire to organize our own life and experience, and act coherently with our personal sense of self. Relatedness is linked to our need to be close and connected to others. The theory stresses that these needs "specify innate psychological nutriments that are essential for ongoing psychological growth, integrity, and well-being"; "it is part of the adaptive design of the human organism to engage interesting activities, to exercise capacities, to pursue connectedness in social groups, and to integrate intrapsychic and interpersonal experiences into a relative unity" (Deci & Ryan, 2000, p. 229). Building on SDT, Oulasvirta and Blom (2008) claim that people engage in technical adaptations that modify the system features because this fulfills their competence (effectiveness of user's actions when acting through the technology) and autonomy needs (transforming a one-size-fits-all technology into a personal belonging).

Implications for the Appropriation Matter of Interest

As humans have fundamental psychological needs and digital technologies are part of the substrates that people use to fulfill them, some uses and appropriations cannot be understood if these needs are not considered. Moreover, and as illustrated above, these high-level psychological needs lead people to engage in activities conducted in their own right and/or intertwined with the achievement of other tasks. Therefore, they can and must be addressed within the activity-centric analytical perspective adopted since the first pages of this book. They are not just influencing factors.

Identifying if and how such needs shape or influence uses raises a specific difficulty: Given the gap between high-level motives (e.g., the need for autonomy or for presenting our self) and the situated motives driving particular uses of applications (e.g., achieving a professional task or having fun with friends), how these high-level motives generate specific activities and/ or impact task-level activities is likely to be largely unconscious. (Indeed, this is also why such analyses are highly important, for both external analysts and individuals reflecting on their practices and their "real" motives.)

One way to address this difficulty is to adopt the top-down strategy illustrated above with SDT: Consider the theories pertaining to the fundamental needs of humans and study how their findings may help understand some uses. The next sections apply this strategy to existential needs, psychological attachments, and engagement.

Before I attempt to make some first steps in this direction, I would like to prevent two possible misleading interpretations of the points made in this introductory section. First, as mentioned in chapter 1, the adopted perspective is not a claim that individuals have absolute agency, and that their wills are the sole explanation and driving force of how they use technologies. The impact of social, cultural, economical, or occupational forces discards this oversimplistic point of view. Analyzing what people do in terms of activities and motivations is not to be confused with the claim that they are, or should be, entrepreneurs of themselves and uniquely responsible for their own destiny. Second, as for the other forces I studied in the preceding chapters, this perspective does not suggest that the general theories studying what we are as humans can *predict* appropriation and/or systematically inform design. The claim is rather that, in certain cases, these theories do help to understand why and how people appropriate technologies in one way or another, and do draw attention to forces that should be kept in mind when designing. For instance, they pertinently call attention to the fact that self-esteem concerns may hinder the use of some features (see the example of word processors and collaborative editing in the next section).

As a matter of fact, some authors have argued that general theories such as SDT can be used to predict adoption and uses. For instance, Kaharana et al. (2018) claim that an SDT analytical lens makes it possible to identify and classify social media affordances but, also, to anticipate the way in which they will be used. For this purpose, the authors reviewed from an SDT perspective different technologies such as blogs, social networks or collaborative projects (e.g., Wikipedia) and then mapped the affordances of these technologies (as identified by the authors) with the needs that they satisfy (again, according to the authors): Egocentric affordances such as presenting oneself, sharing content, or interacting, and allocentric affordances such as forming relationships or browsing other people's content, are categorized as allowing the satisfaction of autonomy needs; affordances for collaborating or competing are categorized as allowing the satisfaction of competence needs; and so on.

From an appropriation perspective, attempting to map psychological needs and technology affordances has similar limits as those of TAM analyses (see section 2.1). For instance, according to Kaharana et al. (2018), "One can predict the psychological needs that motivate use of specific social media applications based on the features they provide. Conversely, given an individual's level of psychological needs, one can predict which affordances and features of a social media application he or she is likely to use" (p. 751). This is not a point I would agree with. As already argued, such perspectives do not acknowledge individuals' specific perception or meaning mechanisms, or the specific system of activities they are engaged in, for example, and are far too positivist. Nevertheless, a priori analyses may inform use studies by putting forward what could be called potential abstract functional values. Excerpts of the authors' analyses that resonate with some of the examples I have used so far in this book include: presenting oneself; indicating one's presence; knowing if others are accessible; forming relationships with others and joining groups or online communities; seeing how others react to one's own presence, profiles, content, or activities; collaborating to create content; and competing with others (Kaharana et al., 2018).

Now that we have clarified these points, let's explore how some general theories may help to decipher certain appropriation cases. The three next sections share the same structure: The theoretical background is followed by illustrations and examples.

5.2   Insights from Experimental Existential Psychology

Theoretical Background

Within the broader field of social psychology, experimental existential psychology (XXP) conducts experimental work studying how people deal with the (inextricably interlinked) existential concerns of death, meaning, isolation, identity, and freedom (Koole et al., 2006; Pyszczynski et al., 2010, 2015b). This approach originally developed from the existentialism philosophical tradition (which, as we will see in section 5.4, also sheds some interesting light on some appropriations of technologies). The study of different issues related to the self, interpersonal relationships, or social cognition led to existential analyses, and the pursuit of scientific inquiry led to the use of experimental methods to study these subjective aspects of human experience. XXP includes different specific perspectives such as terror management theory (TMT; Pyszczynski et al., 2015a) or existential psychotherapy (Yalom, 1980). See Kaptelinin (2018, 2016) for a more general analysis of, and argumentation for, the development of an existential inquiry framework in HCI.

XXP posits that existential anxiety affects or interacts with basic human motives and plays a substantial role in diverse forms of behavior. This anxiety is however largely unconscious, and is managed through the elaboration of social and individual defenses. The understanding of these defenses helps to make sense of human behavior in everyday life (Pyszczynski et al., 2010).

The two core defenses that people develop are (1) maintaining faith in their cultural worldview and (2) living up to the standards of value that are part of this worldview (Pyszczynski et al., 2015a, 2015b). According to some authors, being intimately connected to close others is considered as a distinct third defense or as a means to support the two described previously. As threats to any of these components have the potential of undermining defenses against existential anxiety, humans tend to vigorously defend them.

This, however, may be largely unconscious. XXP has investigated the effects of threats posed by direct confrontation with death and, also, personal uncertainty, meaninglessness, uncontrollability, ostracism, guilt, or shame (Pyszczynski et al., 2015b), and how fluid compensation mechanisms may develop.

While XXP has not (to the best of my knowledge) developed specific analyses of digital technology appropriation or uses, a number of the findings made in this field have some links with the topics we studied in the previous chapters. In the next sections, I propose some tentative analyses of the possible impacts of self-esteem (as addressed in XXP) on the use of digital technologies. XXP defines self-esteem as "the individual's assessment of the extent to which he or she was living up to the standards of value associated with the cultural worldview to which he or she subscribed" (Pyszczynski et al., 2015a, p. 6) and shows how self-esteem may explain social processes such as attitude change, social comparison, stereotype maintenance, or intergroup behavior (Pyszczynski et al., 2010).

Email Overload

How XXP showed that the reflective judgment that one is a valuable contributor to a meaningful world impacts many of our everyday activities sheds some additional light on empirical findings such as, for instance, the fact that some workers fear being judged for their efficacy according to how promptly they manage important messages (Hanrahan et al., 2016). It may be argued that what is at play here is social and/or workplace norms, and not the "real" self. However, as argued by XXP (and SDT), the line between social influence and the authentic self is permeable, and social influences are internalized in the person's core self. This internalization process may be incomplete, in which case people comply with social norms because they feel pressured or guilty, or may be complete, in which case people comply with social norms out of their own volition (Pyszczynski et al., 2010).

With respect to use analyses, there are several interests in identifying if (and, in this case, how) such existential matters are part of the picture. First and foremost, it ensures basic human care. Email overload causes stress and burnouts (Reinke & Chamorro-Premuzic, 2014). In some cases, some workers reach a point where they find it literally impossible to access their emails: Opening the inbox means facing the threat. Secondly, general threats are likely to apply across the board, and an impact on how one task or one technology is addressed draws attention to the potential impact on other cases.

Third, this knowledge can inform remediation processes. As already mentioned, the lack of control resulting from the risk of missing an important email leads some users to avoid using filters and other features designed to screen or reduce email volume (Barley et al., 2011). Attempting to change how these users manage their inbox by raising efficiency objectives would not solve the problem and, actually, may even make the situation worse. What must be identified and addressed is the cause, and not just the symptoms.

Such analyses suggest precise design actions. For instance, in the case of email management: Provide means addressing as such the risk of basic filters erroneously filing out an important message (e.g., via criteria superseding any other filters) and the fact that cutting the technology (here, the email) out of one's life may be impossible and/or raise specific issues. As email is used for different purposes and involves different actors, the inbox may include not only emails that exacerbate the problem but also many others, including those that may be of help. When opening the inbox means facing a threat, it would thus be useful to allow users to set idiosyncratic "existential filters" (e.g., a self-esteem defense filter) that file or delete threatening emails before the users access the inbox messages.

Designing existential filters is indeed a challenging task. However, imperfect solutions may be better than none. Indeed, when developers considered that users needed help to identify the job tasks mentioned in their emails, it led to some attempts to design automated classifiers of emails implementing this goal (Sappelli et al., 2016). An XXP lens suggests the consideration of other types of goals; these are admittedly more complex, and require the involvement of the user him/herself, but are definitely more important.

Social Networks and Experiences Sharing

Similarly, Kaptelinin (2018) pertinently points out that how XXP showed that shared subjective experiences lead to more interpersonal attraction than shared objective characteristics (Pyszczynski et al., 2010) provides both (1) a rationale for the success of social networking applications (in my words, the attribution of functional values to these technologies), which are particularly convenient for sharing subjective experiences (in my words, the features support the perception of their usefulness for such purposes), and (2) a lens to detect how some technical features may raise issues or be incoherent with this activity. For instance, it may be unclear what precise element is "liked" in social media (Is the "like" for the original post, or does it indicate a momentary phenomenological experience related to a comment about the post?).

The importance of this "sharing subjective experiences" motive also sheds some light on the appropriation of communication apps as idiosyncratic communication places, and how it may be responded to via design. The odd yet rational behavior of duplicating resources (e.g., photos) in different silos to maintain independent social spheres could be addressed in terms of the nature of the considered experiences and the people with whom such experiences may be shared. In other words, users could be allowed to customize places rather than applications, and to move a correspondent—with, possibly, the related past interactions—from one place to another (on the same application or to a different one).

Blogs, Websites, and Self-Narratives

Let's now consider Satchell and Dourish's (2009) report of a study where participants were proposed a blog to document their progress in their attempts to stop smoking. When the attempt failed, the blog use fell drastically. Participants' comments included remarks such as "Why would I want to share this with the world?" which leads the authors to point out the role of the technology in how a particular kind of self-presentation is constructed, and how users fulfill responsibilities to themselves and to others. This may be related to an unsurprising finding of XXP: One way for people to maintain self-consistency and the associated feelings of self-esteem is to construct self-narratives (Pyszczynski et al., 2010). This understanding suggests that, when using the blog, users are actually engaged in two intertwined activities. One is to document the efforts to stop smoking and the progress. Another, more general and not specific to this context, is the management (development, maintenance, defense) of self-esteem. As part of the existential efforts for maintaining coherence and security, the latter may indeed supersede the intrinsic (personal) or extrinsic (in response to the experimenters' demand) motivation to document the experience.

This case study once again draws attention to how technical characteristics may support or conflict with such hidden and/or unconscious motives. For instance, the cumulative structure of blogs is very much in line with narratives. However, when an experience takes a bad turn and/or new self-narratives are needed, the blog becomes an obstacle. Basic web pages are much easier to comprehensively change. Typically, refurbishing an old-fashioned web page with recent technical advances is a conscious or unconscious opportunity to update one of the ways we use to address a task that XXP works have identified as core: "Explain to both ourselves and others how the person we were in the past became the person of today, and what path our lives will take in the future" (Pyszczynski et al., 2010, p. 739). Similarly, using an imposed web page format may create issues. It is not infrequent to find personal trajectories and/or experience descriptions in web page slots that were originally intended to enter factual personal characteristics. Collectively deciding on the template used for presenting the members of a group (e.g., for academics, questions such as: Name of the university where the masters or PhD was completed? H-index? Grants? Mentions in general newspapers or TV?) is an exercise that is prone to ontological considerations and elephant-in-the-room situations. It is not surprising that, in many institutions, the chosen template is basic (and includes a link to the external "real" personal web page). Unsurprisingly, empirical analyses show that self-esteem has also a significant effect on perceptions of platform means for identity persistence—that is, "the extent to which a platform affords the identification of content of an individual [amalgamated online 'face'] over time" (de Vito et al., 2017, p. 742).

Word Processors, Collaborative Editing, and Self-Defense

XXP may also help to better understand collaborative editing practices. Studies show how user appropriation of these technologies is impacted by the combined effect of issues such as power dynamics, accountability, and credit of contribution; see a review in Wang et al. (2017). For instance, while comments, track-changes, or revision history dramatically improve collaborative editing efficiency, efficiency is not the only criterion driving uses. Studies show that users are aware of how others will perceive their behaviors that are made explicit by awareness information, and, as a consequence, they take steps to minimize social conflict (Birnholtz & Ibara, 2012; Birnholtz et al., 2013). The XXP perspective suggests extending such analyses to asking if and how awareness features may play a role and/or be impacted by self-esteem maintenance or defense.

As a basic example, one specificity of synchronous collaborative online editors (e.g., Google Docs) is the real-time sharing of updates. A side effect is that it makes editing behaviors (and, thus: first thoughts, hesitations or grammatical errors) directly visible to others. This sheds some light on why some users write their contributions on their local word processor before pasting it in the collaborative document (Wang et al., 2017).

As a more tricky example, let's now study the following fictive but plausible scenario, which many of us may have experienced as the main or a secondary actor. A basic alternative to online collaborative editors is to pass a document back and forth. This type of asynchronous collaborative editing is well supported by the way word processors allow users to add comments and the tracking change process, which highlights the different changes to the text by coeditors. These comments and text modifications are time-stamped, which is necessary for the implementation of the system. However, word processors make these time stamps accessible to users. Although this information may be useful in some cases, it has a substantial side-effect: It literally creates and communicates an implicit activity timespan. Consider a worker, let's say an academic, for whom the self-reflective judgment that he/she is a valuable contributor to a meaningful world partly builds on his/her high professional conscience and efforts to further Science with a capital S. This academic spends six hours reading a printed twelve-page research paper draft and preparing edits/comments in the margins and, finally, enters them in the file. The resulting technological timespan, as revealed by the difference between the first and last time stamps, corresponds to the last phase, which may be only fifteen minutes. And, to add to the disaster, it may be time-stamped at 2 A.M., which some coauthors could consider to be an inappropriate time for dealing with a paper that contains such important and complex ideas and is core to their professional career.

What surfaces here is how technology may lead users to face intertwined and challenging questions related to self-esteem and social relations (as well as those pertaining to authenticity and good/bad faith, see section 5.4). From a reflective perspective, the visible indication of a fifteen-minute working timespan is frustrating for the academics. In terms of being true to oneself, and in view of the fact one is furthering Science, there is no issue. Yet, from a social perspective, there is a risk that others will think he/she has only skim-read the document, and a comment mentioning that this is not the case may not help. It may therefore seem reasonable to avoid this risk of endangering the collaboration, if only for the greater good of producing an excellent paper.

For rhetorical reasons, the presented scenario pinpoints an inaccurate and obviously problematic timespan. However, questions would still remain if this was not the case: Awareness features may also be appropriated as a means to communicate fake information—for example, by skim-reading the paper but spreading a small number of changes/comments over a four-hour period.

In euphemistic terms, it cannot be excluded that such questions impact the way some users appropriate these technologies. It is not surprising that some individuals, once they have understood the side effects of time-stamping, avoid using the collaborative editing means that allow others to "measure" their activity. This leads to workarounds such as disabling the tracking changes mechanism, inserting changes/comments directly in the text and highlighting them with font colors, or an "inadvertent" accept-all-modifications command.

Takeaway for the Appropriation Matter of Interest

The XXP findings, together with those of other approaches such as SDT, strongly suggest that the conscious or unconscious high-level motivations stemming from our human needs may generate specific activities and/or interplay with task-level ones. As we have seen, this understanding helps to make sense of some uses of digital technologies.

The relation linking the use of the technology and the necessity of fulfilling human needs may take different forms. It may be straightforward: As people need to communicate, be part of groups or gain social capital, they are attracted to means that allow them to do so and attribute functional values to them. For instance, although they may not necessarily conceptualize things in this way, many individuals attribute social networking apps a "develop self-esteem" functional value and use them accordingly. However, high-level motivations apply pervasively and may thus also affect or interact with activities driven by other motives. For instance, job-oriented systems of activities may be impacted by same-level or higher-level activities stemming from more fundamental needs (e.g., dealing with existential issues like defending one's self-esteem), which may influence the attribution of functional values to the technologies used in the work practice.

The way the XXP perspective draws attention to high-level and possibly unconscious needs helps to decipher and disentangle the different functional values at play, and informs the search for design options. Typically, in the case of word processors and collaborative editing, a basic customization feature that allows users to avoid time stamps may dramatically change the appropriability of "track-changes" features.

Once again, the point made here is that there are cases where existentially driven motives seem to impact users' appropriation of technologies or, in other words, that some uses are better understood through the application of this complementary analytical lens. For such cases, the specific nature and importance of existential motives calls for very careful attention. XXP claims that many if not most human activities are related to, or impacted by, existentially driven motives (Pyszczynski et al., 2010), which suggests that such motives may directly or indirectly affect many uses of technologies. This remains to be studied further.

As a final word, the examples presented in this section may be subject to different interpretations, the XXP lens being one of them. Linking existential needs and the appropriation of digital technology is thought-provoking, particularly if one relates existential needs to death anxiety. Actually, with respect to the subject of appropriation, these two points may be disentangled. The death awareness explanation is core to XXP or TMT. Other researchers, however, have claimed that the observed effects may be related to other aversive stimuli; see Pyszczynski et al. (2015a, 2015b) for a review of these critics. If, as stated by XXP/TMT opponents, the aforementioned motives do indeed stem from other stimuli, this does not alter the empirical finding that these motives play an important role in human behaviors and, as an output, may impact how we appropriate technologies.

5.3   Insights from Psychological Attachment Theories

My second example of how the theories pertaining to the fundamental needs of humans may help understand how people use technologies is related to psychological attachment, which may take different forms.

Theoretical Background

As raised by Csikszentmihalyi (1993) and many others, our dependence on objects is not only physical but also psychological. Among other functions, including the provision of a feeling of power, things give permanence to meaningful relationships and serve to stabilize and order the mind. As an output, we develop attachments to some of these things and keep them. In Csikszentmihalyi's words, "For most people the home is not just a utilitarian shelter but a repository of things whose familiarity and concreteness help organize the consciousness of their owner, directing it into well-worn grooves" (p. 25).

Different theories address the question of why and how we develop attachments to things. I will refer to two of these, namely the extended self and psychological ownership perspectives.

The notion of extended self (Belk, 1988) builds on the perspective that having, doing, and being are integrally related. Through having and doing, we develop a sense of being, and possessions are a major contributor to, and reflection of, our identities: "Knowingly or unknowingly, intentionally or unintentionally, we regard our possessions as parts of ourselves" (p. 139). Our extended self encompasses persons, places and things, including digital technologies (Belk, 2013, 2016). As these material and immaterial possessions take a role in the construction, maintenance and development of self-identity, we become attached to them. For instance, analyses of archiving practices show that people get attached to and keep physical and digital possessions as traces of personal history and/or social ties (Kirk & Sellen, 2010).

The notion of psychological ownership (Pierce et al., 2003) builds on the perspective that the meanings and emotions associated with the possession of a specific object and the sense that "The object is mine" play a role in intra-individual motives such as efficacy and effectance, self-identity, and having a place to dwell. It embeds cognitive aspects such as the individuals' awareness, thoughts, or beliefs regarding the target of ownership and the emotional or affective sensations they may feel. Here again this perspective acknowledges that, although immaterial, digital technologies meet the characteristics that have been identified as essential for people to develop feelings of ownership (Kirk et al., 2015; Kirk & Swain, 2018) such as being manipulable and controllable, satisfying self-identity motives, and allowing users to develop a sense of place or home.

In order to discuss how these general theories may play a useful role in making sense of certain uses and appropriation of digital technologies, I will present two different points, which I dissociate solely for the sake of explanations: attachment to devices and contents, and attachment to the outputs of appropriation.

Impact of Attachment to, and Ownership of, Devices and Contents

Different studies have provided evidence that digital possessions may be valued in similar ways to their material counterparts when they allow users to express individuality, reflect social ties, connect to the past, or remember loved ones or, in other words, are associated with memories, experiences, emotions, goals, values, and aspects of identity (Orth et al., 2019). Some people become attached to digital devices such as their very first mobile phone or the computer they built from a kit (Kirk & Sellen, 2010). This does not differ from attachment to other material objects. The most common phenomenon is that people become attached to digital contents such as photos, documents, videos, games, or old emails and messages. Kirk et al. (2015) also found that "as consumers successfully appropriate technology, they develop feelings of psychological ownership [with respect to] the people, brands, and products that consumers communicate with through the medium" (p. 166). Examples of behaviors facilitating the emergence of psychological ownership include "composing and publishing self-related content, commenting on others' work, problem-solving in innovation communities, reviewing and evaluating products, seeking and receiving guidance, and other helping-oriented behaviors in user communities" (Kirk & Swain, 2018, p. 77).

These findings help to make sense of use phenomena such as customization practices. Adapting things is both an output and a source of making them part of one's self. Creating or altering things requires cognitive engagement and investment of psychic energy, with an impact on attachment (Belk, 1988). According to Kirk et al. (2015), "It is the level of an individual's perceived influence or contribution [customizing or adapting it to serve new ends] that drives his or her sense of psychological ownership for the technology and the process of using it" (p. 170).

How we develop attachments to things thus provides a background for the argument that nonfunctional adaptations of digital technologies do fulfill goals/needs, although the latter may not be conscious. This is actually part of a pervasive phenomenon: When possible, people tend to contextually and idiosyncratically transform the impersonal artifacts (e.g., cars, clothes, or, in our case, digital technologies) created by designers into their own technology (Oulasvirta & Blom, 2008). Allowing nonfunctional adaptations should thus be an explicit design goal.

Attachment to things raises the question of transferability. With digital technologies, this question may take different realities. An attachment to devices may negatively influence the adoption of applications that cannot be installed on an out-of-date computer or smartphone to which we are attached. The attachment to contents (e.g., photos or documents) does generally not raise any transferability issues. They can be easily copied, and it was found that the attachment to such contents is relatively independent from the host device (Denegri-Knott et al., 2012). For instance, Meschtscherjakov et al. (2014) found that the value attached to smartphones as a means for social interactions is not attributed to the phone but rather to the communication history and/or other symbolic value such as the brand. According to Orth et al. (2019), an account of such phenomena is that the meaning of the technological possession is ascribed at a level of abstraction beyond the singular physical object. An exception, however, is when the application required to access these contents cannot be used on (typically) the new computer or smartphone. Keeping old technologies that we are not emotionally attached to as such, but which continue to provide access to emotionally important contents, is a rational and frequent behavior. This may have an impact on the success or failure of new technologies to enter the ecology of artifacts and be appropriated by users. Techniques such as emulators more or less permit such issues to be addressed. Finally, applications cannot be technically transferred. Users must install the same application on the new device and/or operating system, which is not always possible for genuine (e.g., the product is no longer available) and/or artificially created technical reasons (commercial strategies).

With respect to appropriation, the nontransferability of applications raises the issue of customizations. From a technical perspective, some applications (e.g., email client or web navigators) manage some of the users' personal settings via profile files. Copying the profile from the old to the new device straightforwardly transfers some aspects of how we made the application our own. However, with the exception of very specific cases, the customizations cannot be transferred. We already saw a variant of this issue with Griggio et al.'s (2019) empirical analysis of how some users experience "frustrations around barriers to transferring personal forms of expression across [communication] apps" (p. 26).

What surfaces here is a phenomenon that I argue should be considered, namely becoming attached to the outputs of appropriation rather than to the devices or contents as such. Let's explore this point in further detail.

Impact of Attachment to, and Ownership of, the Outputs of Appropriation

Attachment to the outputs of appropriation may correspond to different nonexclusive realities. The first is related to the technical artifacts. The second is related to the set of uses and associated emotional or psychological constructions that have developed while using these artifacts.

Technical artifacts   One of the findings of psychological ownership studies is that authentic pride—that is, pride as a positive outcome of a specific
behavior (as opposed to hubristic pride and a perception of oneself as supe-
rior to others)—is an important driver of ownership, and that customization of technology produces such feelings of accomplishment (Kirk et al.,
2015). According to Oulasvirta and Blom (2008), "At its best, personalisation becomes rewarding activity in itself regardless of the achieved effects, for
example when personalisable features participate in flow experiences" (p. 1).
Mugge et al. (2009) note that "the effort invested during the personalisation
process has a direct effect (as a result of the extended period of time spent
with the product) and an indirect effect (via the product's self-expressive
value) on the strength of the emotional bond with a product" (p. 469).

This general phenomenon applies to our digital technologies. Our computer or smartphone is often an idiosyncratic resource that we have created
by installing specific applications, making technical or semantic connections
(e.g., relating email and document folders or tags), or changing the neutral
and undersophisticated skins and backgrounds of applications. Moreover,
many users customize and/or enhance the applications they use via plug-ins
(see the next chapter). These are outputs one may legitimately be proud of
and attached to.

When considering new technologies, the fact that they offer innovative or
enhanced features, or are more powerful, is thus just one side of the coin. Our
investment in the shaping of the technologies we have used for months or
years and the subsequent ownership feelings we have developed may make
us reluctant to see them removed from our ecology of artifacts and being
replaced by another one.

As a matter of fact, skilled users can use some dedicated means to list the
applications of the former device and more or less automatize their installation on the new one. In some sense, part of the output of the efforts that led
to the idiosyncratic set of applications forming "our device" may be transferred. This possibility is far from anecdotal for those who, coming back to
Csikszentmihalyi's words, see their computer or smartphones as "a repository
of things whose familiarity and concreteness help organize the consciousness
of their owner, directing it into well-worn grooves." It should be made both
possible and easy for all users.

Emotional and psychological constructions   Another form of attachment
is attachment to the emotional or psychological constructions that have

developed while using some artifacts. This case is illustrated well by the study reported in (Ambe et al., 2017). Two ordinary kettles are augmented with sensing and communication capabilities and placed in geographically distant homes. Using the kettle to boil water in one home lights a lamp in the other home and creates an opportunity to use the communication means (a tablet) associated to the kettles. The experiments reveal how, through its use over time in a family setting, the glowing lamp becomes a symbolic representation of, and initiates thoughts toward, the loved one. As described by one user, "I now associate The Kettle to my mum. When I go to the kitchen, there's fondness when I see the kettle mate glow. . . . When I see it I imagine her in her house in the UK. I know just what her kitchen looks like and I imagine her there" (Ambe et al., 2017, p. 6640).

As highlighted by the authors, this example illustrates how a technology may become distinctly personal because it embodies personal history and emotional attachment. Interestingly, the authors highlight that an important dimension of the development of the technology meanings and emotional valence is the way it becomes part of everyday life. In their words, "Routines and habits highlight the importance of particular objects used within them that support independence, agency, and connection to friends and services. These practices give meaning to our lives, keep us functional and they are cues that everything is in order" (p. 6633).

What surfaces here is that, among others, the technological setting is attributed the functional value of reifying and sustaining emotional bonds, and the properties that implement this value are not easily transferable. The configuration includes the devices (the kettle, the tablet) and the services offered by the applications. However, and very importantly, it also includes the long timespan during which the setting has been part of everyday routine, and the memories associated with it.

Our devices and our applications have an undeniable history. We have shared our lives with them for a significant period of time and, in some cases, have dedicated some efforts to technical modifications. We have also developed associated ways of doing. In other words, our appropriated technologies are the results of a long timespan development process, the engagement of psychic energy (to use Belk's words), important amounts of time and attention, and possibly tricky technical actions. Last but not least, the overall output extends our capabilities and efficiency (as we perceive it). All these elements echo the routes to attachment highlighted by the

aforementioned theories, namely a contribution to the construction, maintenance, or development of self-identity; having a place; intimate knowing;
investment of self; feelings of control, efficacy, and effectance.

Takeaway for the Appropriation Matter of Interest

The general takeaway of this section is that users may develop attachment
to, and/or psychological ownership of, digital technologies (e.g., devices or
applications) and digital contents (e.g., photos). However, this may also be
the case for the habits and uses that developed around and/or resulted from
the technology. As the material or immaterial objects people are psychologically attached to form a "home base," a familiar place providing personal
security (Pierce et al., 2003), abandoning what is considered to be a possession may be difficult and painful, in particular when this possession is the
result of a long-term orientation. People tend to preserve their relationships
with the things they have created and adopt protective behaviors. In their
empirical analyses of why people preserve some things and discard others,
Odom et al. (2009) found that "an object's symbolism . . . can engender a
high strength of attachment when it arises from personal history as a by
product of use over time or when it arises from augmentation that reflects
back on its owner in a personal way" (p. 1061). Alongside other reasons
such as inertia, fear, or cognitive issues, functional values such as "support
memories" may be a factor of resistance to change.

These points allow for a better understanding of some of the developmental aspects of appropriation. The previous chapters showed the functional
impact of the technical and psychological developments associated with
appropriation—that is, how they affect activities such as doing one's tasks
in the workplace or interacting with others. Here, we see that these developments also have an emotional impact. As we construct instruments—that
is, technologies plus ways of doing—we become attached to some of them
for this very reason and not only for their functional benefits. This is particularly the case when technical or psychological constructions extend the
self and/or allow us to overcome difficulties in a way we may be proud of.

This suggests that the emotional valence that results from appropriation may, together with functional aspects, play a role in the adoption and
appropriation of new means. Changes related to the applications and/or
the context within which they are used may positively or negatively impact
functional dimensions (functional values, activity and use schemes) but also

emotional bonds and ownership feelings. What the external observer can consider to be a detail with no functional impact may actually be of great relevance for the user as an individual. Moreover, it is important to note that this emotional valence may, to some extent, be unconscious and/or considered as a private matter, thus making it difficult for an external observer to identify.

Although the use of metaphors can quickly become a slippery terrain, how the material or immaterial objects to which people are psychologically attached form a familiar place providing personal security echoes the suggestion to consider email or social networks as habitats within which many people spend a lot of time. The idiosyncratic organization of folders and subfolders on our computers, or the channels, threads, and other structuring elements of the communication apps or social networks we use on our smartphones, may be regarded in the same way. These familiar places, which both originate from and support the structuration and management of our everyday activities, help us to, in Csikszentmihalyi's words, "organize our consciousness" (in other words, they have this type of functional value). Changes are thus far from anecdotal.

In the light of the analysis conducted in this section, another of the findings of the psychological ownership research tradition is of specific interest. Empirical studies show that individuals are likely to be positively oriented to changes that are self-initiated (reinforcement of the need for control and efficacy), evolutionary (promotion of the sense of self-continuity), and additive (contribution to the need for control, self-enhancement, and feelings of personal efficacy). In direct contrast, imposed, revolutionary, and/or subtractive changes are likely to be resisted (Pierce et al., 2003). It is thus not surprising that some users resist (or face issues with) mandatory applications, in particular when they require changing practices and/or conceptualizations. Typically, moving directly from a management of to-dos via emails to the use of a task management application, or from email to collaborative applications such as Slack, is more revolutionary than evolutionary, particularly if it is imposed on users.

5.4 Insights from the Works Considering Engagement

My third and final example of how the theories pertaining to the fundamental needs of humans may help us understand how people use technologies is related to engagement and, more precisely, engagement as addressed in existential philosophy.

Theoretical Background

As a philosophical tradition, existentialism states that people are responsible for their own behaviors and the individuals they become, and it studies the implications of this statement (Crowell, 2017). While it does not deny the importance of biology or morals, existentialism posits that a further set of categories, governed by the norm of authenticity, are necessary to grasp the meaning of human existence. Renowned existentialists include Kierkegaard (usually considered a precursor) and Sartre (usually considered to be the most prominent existentialist philosopher), with many others who have defined themselves as existentialists or are considered as such.

I will focus on authenticity and, more specifically, Sartre's original perspective as presented in Sartre (1946). (As a matter of fact, existentialism encompasses many other ideas and addresses many other topics, and the claims made by some authors have sometimes evolved from one period to another). In this well-known presentation and then text, Sartre argues that "existence precedes essence." This admittedly cryptic slogan emphasizes the existence of the individual person as a free and responsible agent. An artifact such as a paper knife (this is the example he takes) is designed from the concept of a paper knife and is designed to serve a specific purpose. Its essence— that is, "the sum of formulae and properties that enable it to be produced and defined" precedes its existence (p. 21). Sartre adds that this is a technical vision of the world. In the case of humans, it may be considered that they possess a universal human nature and all share the same basic qualities: Human essence precedes its existence. However, Sartre's perspective is different: *Existence precedes essence* conveys the idea that people exist first, and only then define themselves. In other words, unlike the paper knife, humans are free and are how they have defined themselves through their acts.

Although Sartre's presentation is provocative and debatable, the idea that people should be authentic and true to their own nature is core to the existentialist perspective. The different existentialist thinkers, however, developed very different analyses and prescriptive implications (Crowell, 2017). For instance, Sartre builds on the aforementioned analysis to argue against determinism, advocate emancipation, suggest that we are what we make of ourselves and that this gives us freedom (but the latter has a cost, which includes anxiety), and claim that authenticity should govern our lives; de Beauvoir turned Sartre's slogan into the famous feminist stance "One is not born, but rather becomes, a woman"; but Kierkegaard developed a very different analysis of what it means to be human. Anyway, as we saw in section 5.2, XXP (which has its roots in existentialist philosophy) showed that however humans define or perceive themselves and whatever they want to be, their existential questionings impact their actions.

In the following sections, I first illustrate how engagement can play a role in the use of digital technologies, then study the issues that may be faced by the individuals I propose to label *existentialist users*.

Engagement, FOSS, and Arrangements

Engagement may shape or influence our use (and nonuse) of technologies in different ways. For instance, among the variety of reasons why people may not use technologies, Satchell and Dourish (2009) identified active resistance—that is, deploying a positive effort to resist a technology. The motivations they put forward for such nonuse include privacy, control over personal information or personal time, and engagement or political stances about corporate or state responsibilities. Engagement may also influence the explicit decision to use certain technologies rather than others. For instance, Choi et al. (2015) found that, in addition to financial considerations, motivations for adopting free and/or open-source software (FOSS) technologies rather than proprietary software (i.e., non-free or closed-source software, with licensing rights) include loyalty, ideology, or identification. It may also suggest specific use patterns that have been exploited for nudging—that is, using design to prompt certain behavior in users. For instance, our "tendency to 'be true to our word' and keep commitments we have made" has been used to reduce the risks of student dropouts in online classes by adding a simple button stating "I've started on this Assignment" (Caraban et al., 2019).

Engagement may be sincere, albeit superficial and/or flexible. A classical pattern is that of a FOSS sympathizer who uses applications such as Libre-Office Writer and accepts to pay the price for his/her engagement (e.g., often facing issues when exchanging files with colleagues using mainstream applications such as Microsoft Word); complains when having to engage in a collaborative task mediated by Google Docs or Google Drive; and, when he/she is obliged to use a specific format and offered the choice between Microsoft Word or LaTeX (a powerful but complex system), promises yet again to develop or reactivate his/her LaTeX skills in order to honor his/her FOSS engagement but, meanwhile, switches to Word. As another example, using voice-based smart assistants raises privacy issues that are real (e.g., the storing of all users' commands) or possible (e.g., the recording of private conversations). Studies show that many users, while aware of these issues, trade off their privacy for convenience and act from a place of resignation, considering that resisting is futile as they have already surrendered control of their personal data to the big technology companies (Lau et al., 2018).

However, engagement may also have deep roots, and the extent to which people stick to their principles may significantly impact their appropriation of technologies. Let's study some such cases.

Existential Users and Authenticity

I label as *existential users* any individuals whose motives and consequent behaviors, including their uses of technologies, are shaped and/or impacted by how they define their lives as human beings.

The existential perspective has put forward that deciding how one does something is an action in itself, and thus an engagement. Authenticity—that is, acting in a way that one decides on and commits to, leaving aside social norms and sanctions (Crowell, 2017), is thus core. Actually, independently from endorsing or even being aware of the existentialist philosophical perspective as such, many users consider that what they do, and consequently how they do it, should be coherent with what they want to be.

Let's come back to the FOSS word processor example. The authenticity matter put forward by existentialism helps us to understand why some users do not compromise their views. For the FOSS users who are committed to the cause, what is at stake is being true to themselves and acting according to their principles. This concern for authenticity is directly related to self-esteem in terms of living up to the standards of value that are part of their worldview. Moreover, it may be enhanced by social concerns if they are socially known as FOSS advocates. Actually, for such users, some situations are deadlocks, for example when some institutions only accept native Microsoft Word format.

Let's now consider the potential tensions that such concerns may raise in the case of communication apps. Empirical analyses reveal that some individuals refuse to use these applications for a variety of reasons such as FOSS concerns or data misuse (Baumer et al., 2013). However, there is an important technical difference between word processors and the majority of current communication apps or social networking applications. A reluctance or an impossibility to use a given word processor may be problematic for the user, but workarounds can be found via export/import or copy/paste features. As we have already seen, communication and social networking applications create specific channels and thus oblige participants to use the same application. For users whose quest for authenticity lead to the nonuse of certain types of applications, conditioning participation to social networks to the use of one specific application may thus conflict with the need for close interpersonal attachments. Of course, social networks are not the only means to develop/maintain attachments. However, these technologies have changed and indeed shape how new groups develop, and they have impacted how most previously existing groups communicate in many societies. The lockdown resulting from the Covid pandemic showed how an external factor may impact other means. Moreover, an intense social pressure applies.

Consider the case of a family where just one parent is concerned about FOSS or privacy concerns, while the other uses a social networking application to share daily events and photos with their adult children who have left home. The technology characteristics create tension between two existential concerns, namely authenticity and isolation. The issue does not arise from a lack of interest in communication technologies, or from philosophical (e.g., Neo-Luddism) or religious (Ems, 2014) reasons to avoid using technology. The question here is the contingent technical characteristic of the applications and how it conflicts with a high-level motive. Here, it may be worth recalling that being excluded from an online group may be perceived as a form of ostracism (what is in question is people's internal representation rather than the objective state of affairs) and that XXP showed that ostracism undermines satisfaction of diverse needs, including self-esteem and meaning (Pyszczynski et al., 2010). As already mentioned, big tech's implicit "If you don't like it, don't use it" is not without issues.

Takeaway for the Appropriation Matter of Interest
Although the existential perspective is very general, and what it reveals about digital technology uses and appropriation is thus also very general, it does have heuristic value. As individuals, people may decide to use technologies (or to use them in one way or another) according to their moral values rather than task-level considerations. Conversely, moral values may make some users reluctant to use certain technologies.

For some users, dimensions such as functional organs, developed ways of doing, transparency, or extended self must thus be examined in the light of what the individual wants to be. In *The Plague*, Camus (1947) refers to engagement in life as follows: "But again and again there comes a time in history when the man who dares to say that two and two make four is punished with death. The schoolteacher is well aware of this. And the question is not one of knowing what punishment or reward attends the making of this calculation. The question is one of knowing whether two and two do make four" (p. 128). How one engages in using technologies is, of course, less important than how one engages in many other aspects of life. Nevertheless, it may be part of, or impacted by, a more general engagement. What may at first seem to be a contingent decision/use may actually be a necessary decision/use because, for this particular human actor, here and now, it is a matter of knowing if two and two make four. This perspective resonates nicely with the perspective of considering human actors rather than human factors (Bannon, 1995).

As a side remark, it may be noted that the subjective perspective to digital technologies that is put forward by the appropriation perspective presented in this book bears similarities to the way Sartre, as a philosopher, addresses essence by highlighting subjectivity and agency. Existentialism, as developed by Sartre (1946), is about how humans build their essence by their acts. Appropriation, as I propose to address it, is about how users build, by their uses, what may be seen as the essence of their applications. This happens in all cases, with unexpected appropriations making the phenomenon all the more salient. Sticking to our usual example and Sartre's line of analysis, the inbox of Thunderbird, Zimbra, or Outlook emailing applications is an artifact that is technically designed as an instance of the email-client-inbox technical concept. Nevertheless, for users who appropriated it as a to-do list and use it transparently in this way, the qualities that define the inbox are no longer limited to visualizing received emails: They include remembering things to be done. The fact that the inbox was designed for a different purpose does not change the users' perspective that the inbox is (also) a to-do list—that is, what they define it to be through their actions. In some sense, Sartre's slogan also applies here.

5.5 Conclusion

Enlarging the Functional Perspective

Acknowledging that users are first and foremost humans makes it necessary to widen the analysis of users' motives developed so far. The general explanation for appropriation proposed in chapter 3 puts forward that users are engaged in object-oriented activity systems such as accomplishing their work practices (through the use of emails, word processors, or collaborative editors), exchanging with friends and family (through the use of social networks, blogs, or communication apps), or conducting other activities such as buying on the internet via smart speakers or monitoring their health via smartwatches; and, in this context, these users turn artifacts into resources for themselves. The analyses developed in this chapter lead to complete this understanding as follows: Some of the object-oriented activities users are engaged in consciously or unconsciously derive from, and/or interrelate with (systems of) activities that have human nature-based motives such as the quest for competence, autonomy, authenticity, self-esteem, or relatedness.

These high- and lower-level systems of activities involved interplay in different ways and, in particular, high-level activities may generate new low-level activities and/or impact preexisting low-level activities. An example of activity *generation* is how high-level needs/activities such as relatedness, self-expression, or socially reinforcing one's worldview produce lower-level activities such as sharing experiences via social networks or customizing products. The high- and low-level motives are directly related: Sharing experiences via social networks apps *is a way to* be related to others, to express oneself, and, via the likes that posts attract, to strengthen or justify one's worldview. Examples of *impacts* include email overload or collaborative editing. Here, the motives at play differ between the high-level (e.g., managing anxiety or preserving self-esteem) and the low-level (responding to emails or collaboratively editing texts). The use of the technology is driven by the task level and the necessity of achieving one's job and is impacted by high-level forces such as self-esteem issues. In other words, it is not the low-level tasks in themselves that enable us to preserve self-esteem, but rather the way in which we approach them.

Another way to contrast these different cases is to consider the functional values at play. It may be said that social networks are given a task-level "share experiences" functional value and, taking a more abstract perspective, a high-level "express myself" and/or "strengthen my worldview" functional values. Similarly, a particular device, application, or way of doing can be considered to have a specific "support memories" or "allow me to address the task whilst retaining my cherished artifacts" functional value. However, it would be odd to state that the word processor is attributed a "manage my self-esteem" functional value. This functional value is attributed to the way of using the word processor features—for example, using the color-highlighting features to mark changes or comments while avoiding time stamps (which may be seen as a new hybrid genre solving the privacy issue).

Impact on Use Analyses

It is of course impossible to identify the full set of high-level motives/needs that can possibly play a role in how people use and appropriate technologies. This does not mean that there is no interest in drawing attention to some of them—for example, as a synthesis of the examples reviewed in this chapter: desires and needs such as aesthetics, the quest for competence or autonomy, the reflection of one's self-identity, the maintenance of self-esteem, or the avoidance of anxiety and isolation; phenomena such as the attachment to, and/or psychological ownership of, the artifacts (devices, content) or the habits and uses that developed around the technology (emotional bonds); or voluntary engagements such as the quest for authenticity.

When conducting use analyses, there are many advantages to considering these potential motives/needs as points of attention.

First, in the cases where high-level forces straightforwardly generate low-level activities, widening the perspective provides a better understanding of what is at play, which may be useful to understand how attributed low-level functional values move from one artifact to another: The high-level (i.e., general) activity/need is stable, and what evolves is the implementation (the low-level activity, e.g., sharing experiences *via* social networks).

Second, in some cases it is very obvious how high-level forces impact low-level activities, and we cannot understand how uses unfold if we ignore these forces. The examples I used in this chapter were selected to highlight this point. However, like the example of using inboxes as to-do lists, these examples are likely to be striking and fairly easily identifiable cases of a more pervasive phenomenon. This point may be of specific interest when considering worker wellness and/or efficiency. As we saw, self-esteem or authenticity may lead users to consciously or unconsciously use technologies in a way that impairs their efficiency. This echoes the XXP finding that, when people's self-esteem is insecure or threatened, they struggle to defend or reaffirm it, and the pursuit of self-esteem may paradoxically undermine optimal functioning (Pyszczynski et al., 2010). More precisely, widening the analysis of users' motives provides evidence for the point made in chapter 2: One of the reasons why users' perception of their efficiency may be very idiosyncratic is that they may be simultaneously addressing several conscious and/or unconscious high-level tasks. They consequently consider efficiency with respect to a set of motives that is much larger than anticipated, with these motives differing in nature and in importance.

Finally, given that they are likely to transversely apply to the users' ecologies of artifacts, considering these high-level needs/motives as such may help to find a common rationale for apparently unrelated behaviors. It is easy to identify how the quest for authenticity may lead to FOSS concerns for a variety of applications. However, this may not be the case of self-esteem, which is likely to have a different impact from one application to another. For analysts (or users) attempting to understand (their) uses, identifying that self-esteem plays a role for some applications and/or in some settings suggests that a systematic analysis may be useful. Taking holistic theoretical perspectives may also help to identify how different needs/motives (e.g., authenticity, worldview coherency, self-esteem, anxiety, or attachment) interrelate and/ or interplay.

Some of these high-level motives/needs (or closely related ones) have also been studied through other analytical perspectives than those I reviewed. For instance, Carter et al. (2020) argue that the use of digital technologies may be best understood when considering the notion of IT identity—that is, "the extent to which a person views the use of a hardware device, software application, or software application environment as integral to his or her sense of self" (p. 1315). Criteria include dependence (being reliant or being dependent), emotional energy (e.g., feeling energized or confident), or relatedness (e.g., being linked or close to the technology). Craig et al. (2019) used this perspective to study how resistance to the use of technologies may be related to IT identity threats. The measurement items they propose include loss of worth-based self-esteem (e.g., feeling less respected or admired), competence-based self-esteem (e.g., feeling less confident about having the skills needed to get work done), or authenticity-based self-esteem (e.g., feeling less like the person one wants to be). As one can see, such perspectives may complement and/or intersect those we have studied in greater detail.

Limits and Difficulties

Let's first recall that, except for very specific cases, the rationale for considering the general forces studied in this chapter is to draw attention to their possible impacts rather than to predict specific usages. These forces are broad and apply pervasively, while activity is always situated. The fact that people tend to defend their self-esteem does not predict how they will use social networks or word processors but raises the question of if and how the combination of some uses and some features may lead to situations where self-esteem comes in the picture.

When conducting such analyses, the risks and limits include remaining at a too abstract level (considering users and their needs in general, rather than one user in a particular setting), over-weighting high-level motives (e.g., autonomy or self defense) to the detriment of work practice or leisure contextual characteristics, and/or overinterpreting motives—for example, identifying so-called existential rationales without sufficient evidence.

Moreover, the broadness and multiplicity of the forces that may play a role raises the issue of understanding what happens when values conflict. For instance, we saw how an app (e.g., WhatsApp) may be perceived as a means that allows a user to be in contact with the family (task-level functional value generated by a high-level need) yet conflicts with engagement in FOSS, thus creating a deadlock.

One of the reasons why such settings may reveal particularly complex is the possibly fluid interplay between values. For instance, the LibreOffice word processor may be appropriated as a way to conduct editing tasks while being authentic to one's FOSS engagement. When prompted to use other means, some users compromise and skip to Microsoft Word. However, others may prioritize their engagement motive and, actually, use the situation to contextually attribute LibreOffice the functional value of putting forward their authenticity and their sense of coherence. The central activity changes, the editing concern becomes less important than that of self-presentation, and the involved functional values change accordingly. The blog used to document attempts to stop smoking is another example of fluidity. The sharing experience and enhancing/preserving self-esteem motivations both initially lead to use of the blog. However, they clash when the experience turns out to be unsuccessful, and the (more general and more important) motivation to preserve self-esteem is given priority.

In a nutshell, the key takeaways of this chapter include:

• The functional values we attribute to technologies may be impacted and, in some cases, generated by high-level motives/needs stemming from our human nature. Examples include aesthetics, maintenance of self-esteem, psychological ownership, or engagement.

- When conducting use analyses, these potential motives/needs should be considered as points of attention.
- Given that the involved forces are broad and apply pervasively, users may be largely unconscious of this impact (and/or reluctant to explicitly consider the motives/needs at play).
- The theories pertaining to the fundamental needs of humans may help to identify motives/needs prone to play a role.

Chapter 3 proposed a general understanding of appropriation. Chapters 4 and 5 proposed complements and clarifications that among other topics, have thrown light on why appropriation is a coadaptive phenomenon by specifying the human side of this adaptation (psychological mechanisms, potentially impacting forces). Let's now address the technological dimension of this adaptation (chapters 6 and 7).

6 Appropriation and Adaptation Techniques

Appropriation is a coadaptive phenomenon, and we are active, rather than passive, recipients of technologies: Many users adapt the technologies they use according to the needs of their individual or group practices. The psychological understanding of human development proposed in chapters 3 and 4 proposes a framework to make sense of the psychological dimension of this phenomenon. Let's now consider the technological dimension.

The adaptation means proposed by most technologies originate from the properties of these technologies and their easily modifiable aspects rather than any particular understanding of the phenomena underlying appropriation. Nevertheless, some of these means are congruent with the understanding of appropriation proposed in this book and do allow users to pertinently align practices and technologies.

This chapter takes a technical perspective to different everyday technologies, reviewing and analyzing different examples and forms of adaptations in the light of the analyses carried out in the preceding chapters. This includes means for adapting applications to personal conceptualizations such as organizing resources, using and defining formatting styles, or defining semantic tagging systems (section 6.1); means for customizing interfaces (section 6.2); means for adapting or defining specific behaviors such as deciding what application should be used, automating groups of commands, or taking benefits of automated learning features (section 6.3); means for enhancing applications (section 6.4); means for associating devices with specific actions and automating actions (section 6.5); and means for end-user programming (section 6.6). The objective is (1) to show that many basic adaptation practices actually correspond to appropriation-driven actions and (2) to illustrate some of the issues of presently offered means before studying in chapter 7 how the understanding of appropriation proposed in chapters 3–5 may be used to inform better designs.

To introduce this review, the two following sections provide a brief overview of the adaptation topic (or, as it is usually referred to in computer science, the end-user development field) and then a clarification of the specific perspective on adaptation to which an appropriation perspective leads.

Adaptation techniques Computer scientists have studied how end users can be empowered to adapt software for decades. Many terms have been used to describe this process, such as adaptation, tailoring, and customization. In computer science, the generic term *end-user development* has been defined as the "methods, techniques, and tools that allow users of software systems, who are acting as non-professional software developers, at some point to create, modify or extend a software artifact" (Lieberman et al., 2006, p. 2). I will maintain the *adaptation* label.

This research tradition takes its origins in the fact that user needs are difficult to identify precisely in advance. Providing users with the means to adapt applications to their actual uses is a smart way to address this issue. However, it is indeed fair to say that adaptation is also an interesting computer science research topic per se, and that some works have been driven by the exploration of innovative ideas rather than by users' needs or capabilities.

Different works have dissociated approaches according to the technical nature of the adaptation actions. See Ko et al. (2011) or Barricelli et al. (2019) for syntheses. As an early illustration, Mørch (1997) dissociates customization (modifying the system by choosing its attribute values from a predefined set), integration (adding new features to the system by linking predefined components together with an integration language), and extension (adding new code to the application). Customization includes how the parameters specifying the application interfaces and/or part of its behavior may be defined, typically via a "preferences" menu. Integration is now very common and, in most cases, does not require the use of any integration language: Add-ons and plug-ins can easily be downloaded and integrated. Extension or end-user programming—that is, "programming to achieve the result of a program primarily for personal, rather [than] public, use" (Ko et al., 2011, p. 4)—has benefited from the progress of computer science with respect to both graphical interfaces and software structure (e.g., how component-based approaches and application programming interfaces facilitate the extension or interoperation of pieces of code). This allowed the development of high-level means such as visual programming (i.e., creating programs via a graphical manipulation of components) or programming by demonstration/example (i.e., creating programs via the specification of their outputs). Some authors have argued that facilitating end-user programming may make ordinary citizens more knowledgeable about "the many decisions a democratic society faces about the use of computers, including difficult issues of privacy, freedom of speech, and civil liberties" (Nardi, 1993, p. 3). End-user programming, however, raises important issues such as the gap between the concepts considered by users when using the system and the concepts underlying the system implementation (Stevens et al., 2006) or the necessity to address infrastructural issues (Wulf et al., 2008).

Different classifications of users have also been proposed, for example by differentiating between regular users (who simply want to accomplish their required tasks), customizers (who may customize the system, but do not do any programming), super users (who are skilled with computers, and are interested in exploring tools for tailoring if time permits), power users and local developers (who know how to program using high-level languages), and professional developers (Mørch, 2011).

Appropriation and adaptation   In this book I address users as individuals whose professional and/or personal tasks are mediated by digital technologies and, in this context, may have a pragmatic interest in adapting the systems they use. In line with the understanding proposed in chapters 3 and 4, I thus consider users' adaptations of technologies as a means and possible output (and, thus, possible symptom) of appropriation, albeit a contingent aspect thereof. User evolution is intrinsic to appropriation. Technical adaptations, if any, occur in relation to the user's ways of doing/using and are driven by them. And, as already highlighted by Nardi (1993), end users are not to be considered as unskilled programmers: They are not programmers at all.

Although considering technical adaptations as a contingent aspect of appropriation differs from usual definitions of appropriation, and particularly that proposed by Dourish ("Appropriation is the process by which people adopt and adapt technologies, fitting them into their working practices"), the idea underlying the latter comes close to the overall perspective proposed in this book. Dourish's (2003) argument for considering that adaptation is intrinsic to appropriation builds on the opposition of technologies acting as mechanisms and technologies acting as media in collaborative work. Typically, a workflow acts as a mechanism: It structures users' processes. In direct contrast, email, Slack, or similar communication means act as media: They give users the freedom to define the structure of their collaborative activities. Building on this distinction, "customisation is not simply a method for individuals to adapt technology to meet their own needs; it is, fundamentally, a means by which users can construct their working patterns, individually or as groups, from the basic materials provided" (Bentley & Dourish, 1995, p. 147). This leads Dourish (2003) to point out that, as action is always situated and working practices are adaptive and evolutive, customization is thus inherent to collaborative activity. I would replace the word *customization* in the previous sentence by *evolution of the couple user + technology*, thus including the evolution of the user (psychological constructions), and I would also enlarge the perspective to both individual and collective activities.

In the following sections I study different adaptation means in the light of this perspective, considering these means in terms of mediation: The key issue is whether the adaptations allow users to make the technology (the media) congruent with their conceptualizations, ways of doing, needs, and/ or concerns. This involves considering means that, although technically trivial, are of core importance.

6.1   Adapting Applications to Personal Conceptualizations

Application features and interfaces are based on specific notions and their predefined semantics. In other words, they create and impose the specific conceptualization (the specific differential system of notions) via which users are expected to act. In some cases, however, this may be adapted or extended. This is far from being insignificant. As we saw in chapter 4, conceptualizations play a key role in the perception mechanisms and the achievement of activities, and may evolve along with the technologies appropriation.

In the following, I describe three examples of means for adapting applications to personal conceptualizations (organizing resources such as files, mails, or channels; adapting and defining formatting styles; defining semantic tagging systems) and how they relate to the important points made in the preceding chapters.

Organizing Resources (Files, Mails, Channels)

Let's start by a basic example related to how computer operating systems come with predefined documents, images, and music folders. This organization creates de facto a predefined personal information management means featuring an organization by type of content. However, from a technical perspective, a folder is an asemantic structure: The label is solely a way to identify the folder, which may contain different types of files. Actually, most users do not use the predefined high-level organization by formats. They create their own (more or less messy and/or respected) idiosyncratic organization of folders in the "Documents" folder, when not on the desktop. As a consequence, the overall folders organization and/or their internal structure is highly ad hoc.

What surfaces here is that although it is so basic and usual that we do not even consider this as an adaptation means, the fact that we can define, organize, and label our structuring devices (here, folders) makes it possible to create and use idiosyncratic conceptualizations. Consider the "Holidays" folder that most of us have probably created. This folder can be organized in many different ways, for example by year, by place, by theme, or by traveling companions. Transverse categories may also be used, such as a dedicated "Family" folder. Hopefully, we can define and use the notions/folders that we prefer.

The analyses developed in the preceding chapters allow us to better understand the importance of this basic "defining, organizing, and labeling folders" adaptation means. Let's focus on documents, emails, and their relations.

With regard to documents, Whitham and Cruickshank (2017) found that the manual organization of files contributes to workers' ability to control their environment and, in turn, plays a role in their thinking and working processes. In particular, browsing the folder/file organization is a means for some workers to remember the activity to which the files and folder are related: The absence of folder adaptation means would thus actually compromise users' activities. How folders act as a conceptual means for organizing data may be straightforwardly related to the perception issues addressed in section 2.4: When we want to store a file somewhere, the folder organization provides specific affordances that our perception mechanisms will match with the current need.

With respect to emails, Bellotti et al. (2005) and many other studies have shown how folders are used as a means to implement different strategies such as moving all to-do emails to a specific to-do folder or, in direct contrast, putting all emails except the to-do emails in thematic folders. Making the to-do folder the first of the list by naming it something like "AAA-my-to-do-list" is one of different possible means to attract our attention to it. (As a matter of fact, using a dedicated to-do folder has a substantial cognitive cost, and many users abandon this strategy [Whittaker, 2005].)

With respect to documents/emails relations, Pucihar et al.'s (2016) finding that the names used in the documents files and the email folder hierarchies often overlap illustrates well the need for an ecology of artifacts perspective. The aforementioned study shows that the conceptualization (the different notions) that users adopt or shape to address their activities (e.g., "work" or "holidays") drives the use of the operating system to create a specific folder organization, the use of the email client to create folders or tags, and possibly other applications uses. This conceptualization may originate from the use of one digital or nondigital artifact before contaminating the others, or emerge from the articulated use of the different technologies. A holistic perspective is therefore necessary to understand what is happening.

These different studies highlight the important role of folders as a means to semantically group and then manage different contents, and thus the interest of how some email clients or operating systems allow users to create *virtual folders*. Folders are like boxes: When we store an email or a document in a folder (i.e., associate it with this folder semantic), we are faced with the possible disadvantages of it not being located in another folder (i.e., associated with another semantic) unless we duplicate it, which may raise other issues. Virtual folders dynamically regroup the emails or documents that match specific metadata or keywords. The advantage is that emails or documents may thus be simultaneously part of several virtual folders. For instance, an email may simultaneously appear in the virtual folders related to the task at hand, the to-do list and (for instance) financial issues. In other words, we can simultaneously use different conceptualizations and different nonexclusive ways of addressing items.

As a final example, the way that Slack-like collaborative applications allow/require work groups to define the channels they want to use straightforwardly corresponds to the point made by Dourish that customization is inherent to collaborative activity. The definition of channels dedicated to specific structures, tasks, or matters of interest is technically as trivial as naming a folder. However, doing so allows users to define the communication structure, which is an undeniably important part of how their activities will be mediated. An important difference with the documents or email examples is that here, the definition of channels and the conceptualization of the collective work structure it carries may align with, influence, or conflict with the personal conceptualizations of the group members.

Adapting and Defining Formatting Styles

Similarly to the documents, images, and music folders, the predefined styles offered by word processors and other office applications define a specific conceptualization: They make us think in terms of heading, regular text, reference, bullet-list, and so on. However, the predefined values for font (size, color) or indent parameters of these styles can be changed, and users can also define idiosyncratic styles—that is, define a new style label and associate it with specific parameter values.

Adapting preexisting styles allows the user to attribute his or her own semantics to these labels, and defining new styles permits the creation of new notions such as different styles of references—for example, "things I want to look at later" or "sections of specific interest for my colleague Alice." This is, quite literally, creating a new means to achieve one's tasks.

It may however be noted that the way users are given the opportunity to manage their own styles raises important issues. From a technical perspective, defining or adapting a style requires no more than a basic customization (selection of attribute values from a predefined set). However, styles are based on more than a dozen other notions (e.g., in LibreOffice [2024] word processor: "Position," "Highlighting," "Tabs," "Drop caps," "Area," "Transparency," "Borders," "Font," "Alignment," "Indents & spacing," and so on) that are admittedly difficult to understand. Moreover, styles are organized in a hierarchical structure and inherit some of their properties from other styles. The modification of one style may thus affect others, which is, yet again, complex to understand.

What surfaces once more here is the importance of taking conceptualization issues into consideration. The adaptation means to manage styles is technically easy to use but, as observed for image editors such as Photoshop or Gimp (see section 4.4), its actual use requires the mastering of a complex set of notions, complex interrelations, and related mechanisms. I will present in section 7.2 how the understanding of appropriation as a developmental process proposed in chapter 5 may help address this type of issue.

Defining Semantic Tagging Systems

As a final example of the adaptation of applications to personal conceptualizations, most email clients propose (1) ready-to-use tags (e.g., "to do" or "important") that can be attached to emails and (2) means to create new tags—that is, create personal classification systems. This opportunity may be used in different ways such as to enhance the semantics of the predefined list of tags (e.g., create a "very important" tag), extend this list with new general-purpose tags (e.g., "awaiting third party action") or define tags created for one particular email (e.g., "answer Wednesday after the meeting"). However, it may also be used to create a coherent set of tags serving a particular type of task or activity—for example, "order," "invoice," and "information request" (to take generic labels). In other words, it may be used as a way to import into the email application semantic categories of an external work process and/or genre. Moreover, another interesting features of tags, which echoes the point made on virtual folders, is that they are not exclusive: Emails can be associated to several tags—that is, be addressed via different classification systems (e.g., "order" + "very urgent").

Although defining tags may be considered a similar process to creating/editing styles in a word processor, there is thus an important difference. From a use perspective, the process is similar: There is an existing feature (defining and managing styles, defining and managing tags), and users are allowed to personalize the data (e.g., the names of the styles/tags, the characteristics of the styles, or the color of the tags). However, adapting styles solely facilitates the use of the word processor. In direct contrast, creating a system of tags such as "order," "invoice," and "information request" not only facilitates the management of emails; it also contributes to the achievement of the tasks described in these emails. As for the definition of channels in communication platforms, this stems from the mediating nature of these applications.

6.2   Customizing Interfaces

While most digital technology interfaces provide access to features via predefined menus and/or toolbars, these menus/toolbars are often customizable. This is far from insignificant. The way features are grouped suggests possible actions or semantic proximity. Whatever the predefined organization, it may thus facilitate but also hinder the specific uses developed by individuals. For instance, useless features of overly long menus and submenus may create visual clutter, and frequently used features may require a number of time-consuming actions. Rather than the predefined menus/toolbars implementing the generic one-size-fits-all choice and organization of features as defined by designers, we can define and use a selection of features and an organization that corresponds to and supports our idiosyncratic practices.

6.3   Adapting or Defining Specific Behaviors

Deciding Which Application Should Be Used

From a technical perspective, operating systems allow users to decide which application they want to use to manage their resources. Operating systems associate a given document format with a given application: When the user clicks on a pdf file or a picture, the system launches the application associated with these formats. This association may be customized.

Although customizing the operating system to decide which application we use is, from an emancipatory perspective, a minimum requirement, this basic principle is not granted. Associating a file format (e.g., pdf or jpg) to an application can lead to a long-term battle with an automated reattribution to another application that is more in the interest of the company benefiting from our uses of their application. Unfortunately, the automated (re)attribution process is likely to prevail over the efforts of users to resist over the long term. The analyses developed in the preceding chapters suggest that this is indeed an issue: As we saw, many users just use the applications that are directly available and/or act from a place of resignation.

What this example illustrates is that adaptation is not only a technical issue: It is also a matter of power relations.

Automating Groups of Commands (Email Filtering)

The most classical illustration of user adaptation means is the definition of *macros*—that is, a set of predefined commands. For instance, in a word processor, a macro makes it possible to specify a series of actions to be automatically achieved when a given event such as "open document" occurs. Macros, which are a simple form of end-user programming, literally allow users to extend the set of application features.

Rather than word processor macros, which are rarely used, let's look more closely at email filtering, which is often arguably much more useful. Most email applications allow users to define what actions should be carried out

on reception of certain emails. This is based on interfaces allowing the definition of behavioral rules. For instance, Thunderbird (2024) allows to specify
the targeted emails by selecting an attribute (e.g., "sender," "subject," or
"body"), a property (e.g., "contains," "does not contain," or "begins with"),
and a value. The second part of the filter specifies the action to be taken—for
example, "move to a specific folder" (+ the name of the folder), "tag" (+ the
label of the tag), or "delete."

Filters are a powerful adaptation means: Once one or several filters have
been defined, the email client behaves differently. As a basic example, if some
emails are immediately deleted, the user will not even be aware that he/she
received them. The inbox may also look very different. For instance, consider
the combined effect of defining (1) a set of filters that automatically move the
emails received via mailing-lists to dedicated folders; (2) a set of filters that
tags as "project xxx" any emails where the name of the project xxx appears
and/or are sent by one of its participants; and (3) a final set of filters tagging
as "important" and/or "family" the emails sent by selected persons (e.g., the
manager or our children). Rather than featuring dozens of unread emails,
the inbox will actually feature the emails on which we should focus. More
precisely, it will implement a particular conceptualization of both email and
task management.

As we have seen, using such mechanisms requires the user to be comfortable with automation and its results (remember how the fear of missing an
important email prevents many users from using filters). Moreover, using
such means reifies one's ways of doing, which has its pros and cons. While
making it possible to iteratively refine the implemented process or to share
it with peers, the user's responsibility is technically inscribed in the application in the case of difficulties or failure.

Benefiting from Automated Learning Features

Applications including automated learning features sometimes offer their
users the capacity to act on the learning process. For instance, word processors'
spelling check features are based on a dictionary or, more precisely, a set of
dictionaries for the different supported languages. When an odd wording is
identified, the system proposes possible corrections. Users may however over-
rule the system, state that the word is correct, and add it to a user dictionary.

Taking a mediated activity perspective (see chapter 2), what is at play here
is the user+word processor system. For instance, typing faster than one's own

typing competence results in errors that can be addressed by delegating the task of correcting typos to another part of the system, namely the word processor. This is an efficient strategy: Validating the corrections made by the spelling checker is much faster than taking the time to avoid typos in the first place. And when we tell the spelling checker that "this is not a typo, please add the word to the dictionary," we modify the user+word processor system's overall way of achieving the writing task. Actually, if we are familiar with our recurrent typos or writing habits, we can even define how a word (or an abbreviation) should be corrected every time it appears. In such cases, both the actor+application system as such and its different components develop (coevolution, see chapters 3 and 4): the way the user types the text is modified, and the application (the word processor and its dictionaries) is not the original system anymore but has become the output of the user's appropriation of the original technical proposal. Although the technical modification of the system is basic, the effect on user activity is significant.

This example well illustrates a point made when introducing the notion of functional organs (section 2.3): Considering the user + technology couple as an entity raises the question of which boundary should be considered, the one separating the user and the technology, or the one separating the couple (user + technology) and the outside world.

Another example involving automated learning is the anti-spam feature of email applications. Our passive (e.g., approving or disapproving of it having placed emails in the spam folder) and active (e.g., explicitly defining filters such as "any email from this sender, or using this subject, or mentioning this keyword, should be considered as spam") uses of the anti-spam feature shape the application we use. Two instances of the same email client will actually behave very differently.

An important characteristic of these examples is that users have control, which is not always the case. It is the user who makes the decision to add a word to the dictionary or, to some extent, identify an email as spam. When we use search engines that apply this type of learning mechanism, we are actually also using individualized systems: The information that is retrieved depends on what the system has learned about us. However, we have little control over this.

Here again, emancipatory concerns and power relations are very clear. In terms of the spelling check or spam functionalities, the word processor and the email client are what we want them to be. In direct contrast, the search engine is a combination of what we want it to be (via available customization features) and what another actor wants it to be, namely the company that profits from our uses of the system and sets the learning mechanisms according to its own interests. Unfortunately, the search engine may help the company more than it does the user.

6.4   Enhancing Applications

Plug-ins (also referred to as add-ons or extensions) have become the major way of offering users the capacity to adapt the applications they use. More and more platforms are designed to (1) offer a limited set of core features and (2) provide access to a (growing) set of plug-ins that offer an increasing number of services. For instance, the collaborative platform Slack can be enhanced by myriad services such as getting notifications from other apps (e.g., a drive), starting an external service (e.g., a videoconferencing app), or synchronizing data with other platforms. As an illustration that resonates with our email example, one of these services is a bot that is designed to remember things: The user can tell the bot what should remembered and, when needed, ask the bot using questions in natural language (Wonderbot, 2020). As an illustration that resonates with the need to take an ecology of artifacts perspective, one of the plug-ins can redirect any email sent to a dedicated address to one channel or another.

The understanding of appropriation proposed in this book suggests that, in addition to the fact that their integration is usually technically transparent, the plug-in technique presents several specific interests.

First, plug-ins extend, rather than change, the existing technology. They thus align with the theoretical perspective and the empirical studies having shown that individuals are likely to be positively oriented to changes that are self-initiated, evolutionary rather than revolutionary, and additive (see section 5.3).

Second, the additive nature of plug-ins is a means for designers to respond to uses in two different ways. One is to test how people react and use new features before modifying applications. For instance, one of the most popular Thunderbird email application plug-ins was designed to manage features such as tasks, events, and calendars. This plug-in would later be embedded in the application on the occasion of a new release. Another way to respond to uses, which is particularly in line with the perspective developed in this book, is the automatization of emerging patterns. For instance, the generalization of online repositories of scientific articles leads academics to (1) make extensive internet searches to identify relevant research articles and then (2) connect to the online repositories they know to see if they can find the paper. This pattern was automated by a plug-in that analyses the browsed page, identifies any scientific articles it mentions, searches for them in online repositories, and then, if it finds an online copy, adds a specific link to get direct access from the browser. This type of plug-in does not change the features or look-and-feel of the internet browser as such, but it adds information (a cue that the article is accessible online) and makes a recurrent task easier (the plug-in automates the sequence of actions that is necessary to get a copy of the article).

Last and not least, taking a motivational lens highlights that the installation and use of plug-ins is congruent with one of the native and pervasive high-level tasks rooted in our human nature, namely that of developing ourselves and gaining power. Keeping an eye on the continuously expanding list of plug-ins offered by designers and/or skilled users is, for individuals and groups, an inspiring and practical way to improve their capacity to act.

With respect to this point, examining the plug-ins offered in the context of FOSS applications (e.g., Firefox, Thunderbird, or LibreOffice) reveals that many of them are an output of collective efforts. Individuals build on other individuals' previous designs and/or self-organize as a community of practice—that is, an online group of individuals who share their interest in something they do and in learning how to do it better.

Some companies now also encourage the design of new plug-ins by their users, providing the technical information to do so and the means to collaborate online via forums. Taking a heteromation perspective (see chapter 1), it may be noted that the benefits are not limited to the users. The smart innovative ideas that may emerge cost the company nothing.

6.5 Associating Devices with Specific Actions and Automating Actions

Configurable Internet-of-Things (IoT) push buttons or automation platforms such as IFTTT are good examples of another evolution of the technical ecosystem: offering basic users (meaning non–computer scientists) the capacity to specify how devices can act as controllers of other technologies.

IoT push buttons are wireless tangible devices that can be associated to one or several internet accessible features (e.g., smart home devices that manage the thermostat, turn the light on, or open the rolling shutter). These technologies straightforwardly implement the fact that users literally define what the device is a means for. Similarly, the voice-based intelligent assistants embedded in smart speakers allow to create routines involving different technologies by just telling them, "When I say (x), do (y) and (z)."

Interestingly, the fact that IoT devices can be placed in ad hoc convenient places becomes part of their adaptation features and may play a functional role. As we saw, placing a push button in the corridor leading to the bedroom that is programmed to switch off lights, close shutters, lock the door, and change the thermostat is not only a means to automatize a series of actions; it's also a way to ensure we do not forget to prepare the home for the night.

More generally, an increasing number of technologies (e.g., wireless speakers) are controllable via apps, and installing these apps on a smartphone turns it into an idiosyncratic general remote controller.

With respect to automating actions, platforms such as IFTTT (IFTTT, 2020) allow users to bridge IoT devices and web applications via simple "If trigger then action" rules, triggers and actions being provided by IoT vendors and web service providers. As examples: "If it starts raining, Then close the garage door and turn on the lights"; "If some smoke is detected, Then send an emergency phone call"; or, in line with our usual example: "If an email is starred, Then create an item in the task management app." Such rules may refer to an array of devices such as smart-home devices (e.g., light, thermostat, or camera), wearables (e.g., smartwatches), connected cars, smartphones, cloud storage, online service and content providers, social networking, or communication means such as instant messaging or email (Mi et al., 2017). Analyses suggest that IFTTT-like platforms are rather successful and, once again, highlight the collective dimension, with users defining but also sharing such rules (Ur et al., 2016).

6.6 End-User Programming

Although some positive experiences do exist when specific ad hoc contexts are created, see for example Costabile et al. (2008) or Mørch et al. (2017), end-user programming has not developed. However, there is one success story: spreadsheets.

The success of spreadsheets is an output of different factors. The perspective developed in this book provides a rationale for the following one: In the same way that (re)naming and (re)organizing documents, emails, or folders allows the user to act within a familiar world—one that reflects his/ her personal conceptualizations and habits—spreadsheets allow people to think about their activities in a way that makes sense to them. Early empirical analyses (e.g., Nardi, 1993) already revealed how people easily match spreadsheet features with the specific problems they address. Associating a column, a row, or a cell with task- and/or context-specific labels (e.g., "sales of xxx," "John's number of baby-sitting hours," or "cost per hour") enables users to stick to their conceptualization of the task at hand. In other words, spreadsheets allow users to define and use specific meanings.

6.7 Conclusion

It is often considered that seeking to provide users with some adaptation means (or, in my words, some opportunities to adapt the technologies mediating their actions) is a laudable, but unsuccessful, undertaking. Offering easy-to-use adaptation means has been a hot computer science research topic since the eighties (Paternò & Santoro, 2019). However, with respect to the techniques that computer scientists usually consider as prototypical—that is, macros or end-user programming, the outputs of the efforts made to allow users to significantly modify the application and/or act as designers are far from impressive.

The analysis developed in this chapter reveals a very different reality: Users do actually use the adaption means offered to them as both an output and a means of appropriation. They align the technology with the practice (i.e., modify the mediating substratum) via actions that are supplementary to the basic use (e.g., creating ad hoc folders or email filters) and/or necessary for this basic use (e.g., defining communication channels in Slack). More generally, many people use adapted/appropriated smartphones and computers that they have literally designed in use by installing—or asking relatives to install—an idiosyncratic set of apps, followed by plug-ins.

As for other features, the attribution of functional values to adaptation means is rooted in needs and motivations. The renaming of folders is a good example. Indeed, it would be difficult to use a computer or an email client where all the created folders remain labeled "New Folder." As examples of less instantaneous adaptations, people's motivation to express their self is suffi-cient to make them change the appearance of the applications they use, the background image of the laptop desktop, the ringtone of the smartphone, or the appearance of their avatar (see chapter 5). Nevertheless, complex adap-tations are hindered by a well-identified issue: Needs for adaptation appear while dealing with task-related activities, and adapting the application in this context breaks the flow. Typically, when we write a text with a word processor, our attention is focused on the text. Although we may realize that the edition of the text will be simplified if we change some parameters, we may not want to stop our activity, and actually, this is generally wise behav-ior. Later on time has passed, and we may have little motivation to address the issue.

With respect to the motivation issue, it must be kept in mind that dif-ferent realities coexist. One reality is the way people voluntarily decide to select their favorite digital technologies from an extensive offer to manage their music, buy goods online, or communicate with friends. Another is how workers are obliged to complete specific tasks and use specific applica-tions to achieve them. In the latter case, the implication of not offering per-tinent adaptation means is that the only adjustment variable is the worker him/herself.

Actually, when adaptation means are lacking, motivated users sometimes find innovative ways of aligning the technology with their practices—for example, by cheating the system. For instance, different studies showed that the way enterprise systems impose predefined ways of working and prescribed sequences often hinders users' contextual activities. In order to successfully conduct their work practice, users thus develop workarounds such as using a field to represent another type of information than what it was designed for (i.e., changing its semantics) or entering dummy data to obtain a given behavior. See, for example, Alter (2014).

Finally, analyzing adaptation means and their uses once again highlights that it is very important to take an ecology of artifacts and collective per-spective, particularly given the possible interoperation of Web technologies. Slack's (2020) website claims, "Connect your tools, connect your teams. Slack channels are the place for work. Bring your tools to your team, save time and avoid context-switching." In an interesting study that illustrates the role of collectives, Lambton-Howard et al. (2020) describe several cases of groups creating original ad hoc collaborative platforms by coordinating the use of different applications (e.g., Facebook, WhatsApp, or Instagram) and adopting specific norms or structured behaviors.

In a nutshell, the key takeaways of this chapter include:

- Current basic adaptation techniques permit designers to provide users with useful adaptation means.
- Users do use some of these means as both an output and a means of appropriation. (The claim that users do not use adaptation means is largely related to an inadequate analysis and interpretation of their practices.)
- Designers should carry out a more detailed analysis of why and how users use current adaptation means (see the presented examples for inspiration).
- As for other features, the attribution of functional values to adaptation means is rooted in needs and motivations: What is core is if and how the adaptation means allow users to pertinently align practices and technologies.

As we have seen, while current techniques already permit designers to provide users with useful adaptation means, these techniques are not always optimally exploited by designers, and there is indeed room for improvement. In the next chapter, I explore how the understanding of appropriation proposed in chapters 3–5 may support better designs.

7 Designing for and from Appropriation

Appropriation is a phenomenon. It cannot be designed. It can emerge or not occur at all, and the form it takes is not fully predictable. Designers have limited control over how their applications will be perceived and used. Nevertheless, it is possible to act on the factors playing a role, and these include the characteristics of technologies.

The likelihood of users appropriating a technology may be increased by facilitating the underlying psychological and technical processes. This objective should therefore inform the overall design of applications and their life cycle and, as we saw in chapter 6, the proposed adaptation means. Moreover, in line with an emancipatory perspective, this objective should be addressed by empowering the users rather than acting for them.

The first section of this chapter draws a general picture of what may be referred to as designing *for* and *from* appropriation. Next, I will make specific proposals building on the understanding of appropriation proposed in this book and, in particular, the importance of conceptualizations (sections 7.2–7.5). In view of the native advantages of plug-ins as highlighted in section 6.4, I will focus on this technical approach. As the case of AI-based technologies is very specific, I will study it as such (section 7.6). Finally, I will summarize some relevant complements proposed by other works, including the TEAS approach, semiotics engineering, infrastructuring, automated interface adaptation, and meta-design (section 7.7).

7.1 General Picture

Being Attentive to Users' Actual Needs and Practices
Although very obvious, let's first reiterate the importance of identifying users' actual needs and practices. For these purposes, user-centered design

(UCD) and participatory design (PD) approaches have proven their value. UCD recommends an iterative process where designers focus on the users and their needs, characteristics, environments, and activities. PD recommends involving all the different stakeholders, and particularly the users, in the design process. Such approaches undoubtedly contribute to increase the likelihood that users will appropriate the designed product.

With respect to UCD and PD general principles, the understanding of appropriation proposed in this book calls for adopting the following core principle: The analytical focus must be on *how users interact with the tasks they consider*, rather than on how they interact with the technologies under design (see chapter 3). As we have already seen, this includes seeking out and identifying factors beyond the users' apparent functional needs and generic usability concerns. The AT and genre approaches provide pertinent analysis means. For instance, Bødker and Klokmose (2011) propose considering the three levels of activity theory from both the human and the artifact perspectives. This framework helps to analyze what users do (What do they say they do? What is the artifact used for? What can it be used for?) and why they do it (motivational aspects). This model also makes it possible to study the dynamics of users' ecologies of artifacts (Bødker & Klokmose, 2012). The genre approach suggests analyzing genres and uses of technologies by considering questions such as "why," "what," "who/m," "how," "when," and "where" (Yates & Orlikowski, 2002). As shown by Spinuzzi (2003), taking a genre approach helps identify the tacit—and thus not easily identifiable—structures or rules shaping usages, and can be conducted at the three levels suggested by AT.

Several UCD approaches pay specific attention to the importance of considering users first and foremost as humans. One of these is "reflective design" (Sengers et al., 2005), which argues for acknowledging people's ways of seeing and experiencing the world. Reflective design highlights principles such as "engaging designers to identify unconscious values and assumptions" that are built into design or helping users to "reflect on, and perhaps reject, how our technology is influencing their choice of activities and their engagement in these activities" (p. 55). As another example, the aims of the "designing for the self" approach (Zimmerman, 2009) include "the design of products that help people move closer to their idealized sense of self in a specific role through their interaction with the product" (p. 395). This approach suggests considering aspects such as role engagement, control, affiliation, ability (and bad habits), rituals, or long-term goals. Keinonen (2008) calls for user-centered design to "protect users from harm" and cater to users' fundamental needs, including those of freedom and self-esteem. A more general tradition addressing such concerns is value-sensitive design (VSD; Friedman & Hendry, 2019). The VSD line of thinking is that all technologies both reflect and affect human values. As a consequence, values must be taken into account in the design process, and design may be used to improve the human condition. VSD emphasizes values such as ethics, morality, human well-being, dignity, and justice. The way VSD addresses values, however, is not without practical (and, also, philosophical) issues.

Finally, taking a more technical perspective, the generalization of object-oriented approaches (since the 90s) and the Unified Modeling Language (UML, 2017) has made the consideration of "use cases" a standard practice. Such scenarios are defined by identifying the users, who are referred to as the actors of the system, and describing how they will interact with the system (actions or event steps) to achieve their tasks. Major progress was made by asking the secretary or the accounting officer to depict the course of their actions and their needs, rather than asking their boss or other people who did not carry out the activity themselves. However, these approaches prioritize the roles played by the user, rather than his/her human nature. What is taken into account is how a panel of (for instance) accounting officers describe the actions and needs of an accounting officer. Moreover, the focus is on how they interact with the system rather than how they interact with the task. Agile methodologies (e.g., Scrum), which are presently mainstream, build on a similar frame of mind: Design is organized on the basis of "user stories" that are identified via patterns relating roles and capabilities.

Although care should be taken to avoid throwing the baby out with the bathwater, these approaches remain unsatisfactory: They must be enhanced to address the appropriation phenomena, and in particular the question of inter- and intra-user variability.

Addressing Inter- and Intra-user Variability

As we saw in chapter 3 and studied in closer detail in chapter 4 development (i.e., the fact that we change and adapt across the course of our lives) is an ontological characteristic of humans. The way we conduct activities and use technologies stems from past developments (i.e., the psychological constructions developed to date: notions, more or less patterned ways of addressing tasks and using technologies) but also nurtures the continuing development thereof. This is the rationale for considering the quadruplet (task, way of doing, artifact, way of using) and its stability/evolution: The cognitive structures shaping the patterned way of addressing the task and the cognitive structures shaping the patterned way of using the technology in the context of this task may both evolve, and these evolutions are not necessarily related to an evolution of the task or of the technology.

The implication is that appropriation and, more specifically, its developmental nature, leads to both inter-user and intra-user variability (Tchounikine, 2019a), two issues that traditional design approaches do not consider as such.

First issue: Inter-user variability   *Inter-user variability* refers to how different users may develop different instruments from the same technology and, for this reason, may present slightly different functional needs to develop efficient idiosyncratic resources. The first reason for such inter-user variability is the developmental mechanism recalled above: People develop differently. Another factor for inter-user variability is that people act within idiosyncratic ecologies of artifacts, and the genealogy of these ecologies or the way they evolve as such (e.g., adoption of a new application) or as a system (e.g., modification of how different applications are used in coordination) differs from one individual to another. Although more may be known about the sometimes imposed technological ecosystem in structured work practice settings, we have seen that workers may and often do use unexpected additional means, the personal smartphone being an obvious example. Finally, the specific context within which the technology is used may of course also play a role.

As already highlighted by Spinuzzi (2003), the way UCD fieldwork-to-formalization methods "assumes that researchers can easily move from the particular to the general, from divergent local practices to a single ideal model of the work" is thus an issue. UCD or PD approaches are indeed useful to understand how work practices or general social forces shape or impact users' interactions with tasks. However, their objective is to design applications that are functional for users (in the plural). As discussed in chapter 1 and then section 2.6, acknowledging the collective dimension must not overshadow what happens at the level of the individual. People develop as groups but, first and most, as individuals.

Approaches such as contextual design, workplace ethnography, or other qualitative perspectives for mapping out users' understandings, practices, and resources may be of help to consider some of the aspects put forward in this book. For instance, the contextual design approach (Holtzblatt & Beyer, 2014) specifically attempts to "understand users in order to find out their fundamental intents, desires, and drivers." It acknowledges that the use of technology is always situated in a larger environmental context and that people are not always conscious of their practices, particularly because many aspects are tacit and "work practice [in a general sense, i.e., 'the complex and detailed set of behaviors, attitudes, goals and intents that characterize a set of users in a particular environment'] is complex and varied, and that useful design data are hidden in everyday details." It also underlines that the introduction of new technologies will change the environment. The contextual design approach thus suggests analyzing practices in the users' own work or life environments and proposes to engage users in a reflection on their actions, intents, and values. For this purpose, it suggests modeling not only how users achieve their tasks but also the communication and coordination between people to accomplish their activities, the cultural aspects that may play a role, and so on.

While their general guidelines are pertinent and useful, these approaches do not consider as such the idiosyncratic aspects of mediation and how they are shaped by individual's perceptions, conceptualizations, or usual ways of doing/using (see chapters 3 and 4). As a consequence, they do not specifically consider how this may be acknowledged and responded to by design.

Second issue: Intra-user variability   *Intra-user variability* refers to how one user may develop different instruments from the same technology in the course of his/her activities and, in this context, may demonstrate changing technical needs. Like inter-user variability, intra-user variability may stem from the developmental mechanisms recalled above, the evolution of users' idiosyncratic ecologies of artifacts and/or the evolution of the context.

While inter-user variability is easy to understand (though not necessarily easy to acknowledge in design), intra-user variability is more destabilizing. Nevertheless, whatever the accuracy of needs analyses, one of the consequences of the developmental nature of appropriation is that a technology may be initially perfectly adapted to a user's needs and raise issues at a later date. Although including users in the design (PD) is likely to be beneficial, users' understanding of their practices suffers from the same issue as the designers' understanding: How things will evolve cannot be completely foreseen.

One design is one response and, although it may be based on careful analyses, it remains a one-off response. Therefore, the only way to address inter- and intra-user variability is, in addition to careful analyses of users' (in the plural, i.e., shared) needs and practices, to provide these users with adaptation means they can use as individuals or as groups, and to design these adaptations means in the light of an understanding of appropriation phenomena.

Designing for and from Appropriation

Designing *for* appropriation may be defined as explicitly considering the factors that may enhance the chances that appropriation will develop. For instance, we saw in chapter 6 how the theoretical understanding developed in this book provides a rationale for the general guidelines proposed by Dix (2007) such as providing elements where users can add their own meanings (e.g., coming back to the examples presented in chapter 6, naming and organizing folders or communication channels, or defining idiosyncratic systems of tags), exposing design intentions, and supporting use rather than controlling it.

Designing *for* appropriation is to be coupled with designing *from* appropriation—that is, in a way that is informed by an analysis of the outputs of users' appropriation processes. This echoes the general principle put forward in Carroll's task-artifact cycle (Carroll et al., 1991) but may be reframed using the SER model (seeding, evolutionary growth, and reseeding) proposed in the meta-design manifesto (Fischer et al., 2004): An initial professional development provides a seed plus means to complete or extend the system; during the period of use, users can adapt the system and continue its development (evolutionary growth); and the reseed phase is a professional development that organizes the changes developed during the evolutionary growth. This general principle of alternating professional development (the seed and reseeds) and use periods within which users may adapt and continue developing the application can indeed be found in other end-user development approaches, see, for example, Mørch (2011).

Let's now review a certain number of more precise guidelines, principles, or approaches suggested by the theoretical understanding developed in chapters 3–5 and how they address some of the issues or limits raised in chapter 6. As mentioned above, I will focus on the plug-in technical approach.

7.2 Being Attentive to Conceptual Issues

Using Plug-ins to Address Conceptual Flexibility and Variability

As we saw in chapter 4, people often use imprecise notions and/or adapt them to the setting, which may lead to difficulties when having to use the precise and rigid notions implemented in the technology. For instance, when using Thunderbird or similar email / personal information management applications, a task is to be thought of and represented via a predefined fixed structure featuring fields such as "Location," "Category," or "Due Date" (see section 4.4). Such designs impose *one* conceptualization of the notion of task, which may come into conflict with how users may have different situated needs. As another example, many email applications allow users to create a task from an email via a right click. While this definitely helps to connect email and task notions, it also defines a 1–1 relation between emails and tasks. When the system does not allow to link the email with an already existing task, the 1–1 relation may become an issue for messages addressing several topics, which occurs very frequently (Reyes & Tchounikine, 2003), or when one task is related to several messages.

The improvement that is needed here is to offer some conceptual flexibility, which may be addressed by providing users with basic customization means. In the case of email applications, such customizations may easily be provided via plug-ins (Tchounikine, 2019a). For instance, in addition to the predefined task structure, users could easily be offered a plug-in allowing them to define idiosyncratic variants (e.g., a basic informal task structure such as "name" + "open text description," or, on the contrary, different specific structures including ad hoc slots) or to associate several emails to one task. Such customization means straightforwardly address the fact that one user may have different situated needs (*conceptual flexibility*) and, indeed, that different users may have different needs (*conceptual variability*).

It is fair to say that offering such flexibility is not always possible. The point is not that it should be offered in all cases but that unnecessarily imposing a predefined and/or rigid conceptualization should be avoided, when possible. Typically, there is no technical need to offer just one task structure, whatever the quality of the proposal.

Using Plug-ins to Address ZPD Issues

As we saw in chapter 4, individuals may only use and appropriate means involving notions that are part of their actual practices or, if scaffolded, their zone of proximal development (ZPD). It is thus necessary to be attentive to users' conceptualizations but, also, to how they may evolve.

Let's take as an illustration an individual whose basic word processor practice includes (1) selecting portions of text and applying successive basic formats such as bold and then italics and (2) using the format paintbrush to copy the complex format of a portion of text and apply it ("paste it") to another. Although presenting identical behavioral achievements, some users may address the copy/pasting technique in a purely empirically way (making two portions of texts look similar), when some others may be implicitly aware that "something" has been copied and then pasted. Although not conceptualizing it as such, the latter mobilize the notion of style without being aware of doing so. The respective ZPD of these different users, and thus the way they may evolve thanks to the scaffolding of peers and/or the system interface, may thus be different.

Although measuring individuals' ZPD (which requires a precise analysis of what they can do alone and what they may do with scaffolding) only makes sense for very specific cases, the ZPD notion suggests a useful general guideline: Users act within a given conceptual and technical zone; they may be helped to extend this zone and improve their practices by drawing their attention to some new notions, but presenting too many or too complex (too "far") new notions is useless, if not counterproductive. For instance, as we saw in section 6.1, the notion of "style" is abstract and complex. Most users can only progressively develop a more or less pertinent conceptualization of "style" and the underlying dozen notions—for example, "paragraph property." Similarly, users cannot be expected to straightforwardly understand how styles are organized in a hierarchical structure and inherit some of their properties from other styles.

This guideline resonates with the fact that, as additive technologies, plug-ins avoid the dilemma stemming from the search for a one-size-fits-all-needs design. Designing *one* interface for adapting styles raises the issue of deciding what notions and/or mechanisms it features. Whatever the choice, it will be meaningful and relevant for the users who have developed the involved notions but, for the others, is likely to be cryptic. This is particularly the case when this interface provides the full set of means—in other words, is for expert users.

What surfaces here is that offering several plug-ins that highlight fewer/ simpler notions and/or deal with the complex mechanisms (e.g., property inheritance) without presenting them as such would increase the chances that some of these plug-ins (i.e., some of these adaptation means) will be in the space within which the user may act productively, or in his/her ZPD.

Such a strategy would reflect several of the points made in this book: make the customization means more usable, and thus contribute to the appropriability of word processors; address inter- and intra-user variability; allow users who are gaining expertise to change their practices (via new plug-ins) in an evolutionary rather than revolutionary way; and, in line with the ZPD perspective, help users to close the gap between the notions underlying present practices and those underlying the fully fledged means.

7.3 Drawing Users' Attention to Potentially Useful Means

Another implication of the fact that uses generally develop in an evolutionary and additive rather than revolutionary way is that users who invested time and energy in developing a way of doing and, possibly, using adaptation means to improve the system fit, may be interested by the plug-ins that allow them to continue in what may be labeled an *instrumental direction* (Tchounikine, 2019a). For instance, consider users who have attributed an "ad hoc semantic system" functional value to tags and use them to manage specific tasks (e.g., have created a coherent set of tags such as "order," "invoice," and "information request"; see section 6.1). These users may find an interest in plug-ins that may be attributed a similar functional value—for example, a plug-in designed to structure tags into categories and subcategories (e.g., the different types of orders and/or possible associated actions): When tagging an email, the categories and subcategories will prompt for the different options and bring to mind the overall and then detailed organization one has adopted. This well resonates with how the possibility of renaming and organizing files and folders contributes to the worker's ability to control their environment and, in turn, plays a role in their thinking and working processes (see section 6.1).

A possible strategy for increasing the likelihood that users will draw links between their activities/practices and the plug-in offer or, in other words, perceive some plug-ins as affordances, is thus to adopt an activity-related (rather than topic-related) presentation of these plug-ins and use functional values as indexes. As an illustration: "If you use your email to manage to-do lists, this may be of interest for you," followed by details of the different identified practices such as taking advantage of tags or filters. Using (expected and unexpected) functional values as indexes draws attention to, and legitimates, idiosyncratic practices (what one does, why, and how) and may attract users' attention to means mobilizing other notions than the ones they used so far.

As an illustration, conducting in the light of the "task management" instrumental direction a systematic analysis of the very long and in-expansion list of plug-ins offered for Thunderbird (or any other email client) reveals many potentially useful means that are not presented as tools for this use, differ in nature, and are thus unlikely to be identified by users (Tchounikine, 2019a). This is for instance the case for a plug-in that allows users to specify an action (e.g., display a prompt message) that must be executed when a new mail is found in a given folder. This plug-in may be used to create semantic folders (e.g., one folder per task), then create filters that automatically file emails in folders, and finally associate a prompt with these folders: When a new email associated with the "observed" tasks is received, an alert will be given. As other examples, the plug-in permitting users to plan the automated sending of a message at a given date/time may be used to send themselves a reminder; the plug-in that allows users to add Post-its to messages solves one of the known issues of using emails as reminders, namely the cognitive cost of rebuilding one's thoughts each time one reconsiders the emails or tasks; and the plug-in that allows users to associate an email with a deadline date for responding to it can alternatively be used to mention the deadline for dealing with the task. Such a systematical analysis also reveals gaps and suggests smart enhancements—for example, a plug-in for sending and receiving tags with emails, which would allow one to share the elaborated semantic organization with other users.

The specific interest of considering the abstract level of instrumental directions is that directions may prove more stable than functional values. This point is illustrated well by the social networks turnover. Periodically, studies reveal the evolution of (in my words) the different functional values attributed to WhatsApp, Facebook, Instagram, Twitter, Snapchat, TikTok, and so on. For instance, in Boczkowski et al.'s (2018) study, the specific functional value attributed to Instagram is "careful and stylized constructed visual portraits of everyday life." This functional value may be seen as part of a more general "gain social capital" instrumental direction. While the attribution of the functional value to Instagram may change, typically because it is transferred to another platform, the instrumental direction, which stems from a fundamental need, is likely to be stable for some period of time. When email is an old and enduring technology, the turnover of digital technologies is now much more rapid. However, this is the case of particular implementations rather than paradigms. Digital interconnections and communication protocols advances opened the way for social networks or collaborative platforms, which are indeed very different from email. However, the specific social networks or collaborative platforms that coexist and/or replace one another are all more or less the same. What is crucial and should thus be focused on is the nature of the paradigm and of people's uses as related to their needs and motivations, which is much more stable.

It may be noted that more and more tech companies now provide users with online means to share and exchange experiences and actively encourage them to do so, which resonates well with my proposal to use functional values as indexes. The success of these initiatives provides evidence of the benefit for users. When such sharing dynamics develop, the object of appropriation is in some sense the technology plus the discussions, shared experiences, and reusable materials that develop around it. What has been integrated in the practice is the application and the associated support. Here again, the economic value costs the company nothing.

7.4 Offering Under-Designed Means

The obvious way to involve and support users in considering the activity/technology fit is to engage them in work practice analysis workshops. This may be addressed in the light of approaches such as participative design, meta-design, or, taking a wider scope, methodologies for transforming collective activity systems (Engeström, 2009). Such strategies, however, only make sense for willing users and/or workers who have to engage in such workshops as part of their job.

The understanding that appropriation stems from the constructive nature of activity suggests another option: providing users with under-designed features—that is, a system that is not directly operational for the user's task—and requiring users to "finalize it" in the context of their activity. Although trivial, this is what happens when creating a new folder: The system prompts users to finalize the action by changing the by-default "New Folder" label. Similarly, collaborative platforms require users to define and label the communication channels and thus customize the application. As a more complex example, the learning scenario editor introduced in section 3.2 cannot be used straightforwardly (Sobreira & Tchounikine, 2012, 2015): Differently from other editors, users must first select and order the notions (e.g., students, groups, or learning activities) they want to use. In other words, the teacher, as the owner of the problem (designing and editing a learning scenario), decides on the best means (the best modeling structure).

The notion of under-designed means as I introduce it here, and the fact that users must complete the design, is of course a user perspective. Technically, there is one piece of code, developed by professionals, which is complete and cannot be modified by users. This is not to be confused with under-design settings where an incomplete piece of software is provided to users who then have to complete it by going into the code.

Although the under-design perspective only makes practical sense for some specific cases, it presents several specifically interesting characteristics: the adaptation process that is required to obtain a usable system is a task-related process; the additional task (e.g., defining the name of folders or channels, or the learning scenario representation structure) is part of the overall reflection on both the task at hand (putting together a set of files, elaborating a communication structure, editing a learning scenario) and what is at stake (personal information management, collaborating, teaching); the technology adaptation coincides with, and is a part of, making sense of the setting; finally, it is natively part of the engagement in the task (rather than an additional activity).

7.5 Favoring Productive Analogies

Among other options, the role of patterns in how we perceive and understand the task to be achieved or the technology to be used may be addressed by taking an analogical perspective. In the cognitive psychology discipline, works that study problem-solving have revealed that analogies play a core role. Although, to the best of my knowledge, the link between appropriation and analogies has not yet been explored, the findings of these works can inform design.

According to Hofstadter and Sander (2013), analogies are "the fuel and fire of thinking." (p. 3) They claim that categorization and analogy are actually the two facets of the core mechanism of human development: the creation of a more or less idiosyncratic repertoire of conceptual categories. Individuals constantly and unwillingly/unconsciously evoke categories and use them to make sense of—and act in—known and, also, new situations. Here again, the authors clearly reject a basic classification process based on precise and fixed structures. A situation may and generally does evoke different categories, and categories are to be thought of as fluid and evolutive.

This analogical perspective provides one possible explanation for a mechanism that, as we saw, is core: What is known guides the perception and discovery of the unknown (chapters 2 and 4). With this respect, the different points posited by Hofstadter and Sander resonate well with the analyses developed in the preceding sections: Individuals tend to give priority to usual categories—that is, what they know well; people tend to draw analogies from the properties that appear salient to them, which may be, and often are, the superficial rather than the structural properties; the repertoire of categories and/or the categorization process may lead to idiosyncratic interpretations; and concepts play a key role.

Let's develop a tentative analogical analysis of collaboration platforms such as Slack. Although their promoters claim that the concept is original, users are likely to perceive these platforms according to the categories of communication applications and/or the categories of forms of communication that these users have developed from their uses of previous technologies such as emails, forums, wikis, instant messaging, or chats. The technical characteristics of these tools can lead to different categorizations. One is to dissociate synchronous (e.g., instant messaging or chats) and asynchronous (e.g., email or forum) means. Considering this property may lead a user to perceive Slack as a synchronous means, and it is indeed often presented as a generalized chat. However, this synchronous/asynchronous categorization is partially misleading. In practice, people are hopefully not constantly keeping an eye on the channels. As a consequence, it is often unclear whether an immediate response is required and/or will be provided in time. Another frequent categorization is to dissociate correspondent- and topic-based structures. When using email or chat, one sends a message to an individual or a group of individuals. In direct contrast, posting a message in a forum or contributing to a wiki is a topic-centered process, and the user is not necessarily aware of who will access it. The Slack notion of channel can relate to both topics and groups.

What this analysis suggests is that, due to the role of analogies, individuals may develop divergent interpretations of communication platforms, with an impact on how they perceive new means. For instance, for users who have developed categories that dissociate synchronous and asynchronous means or correspondent- and topic-based structures, the characteristics of channel-based platforms may lead to some confusion and/or unexpected understandings, with an impact on uses. As another illustration, a potentially tricky characteristic of some communication platforms is that previous messages may be edited or deleted. This impacts the very notion of ongoing discussion and may even conflict with perspectives where what has been written may be commented on or enhanced but cannot be changed.

The categories resulting from the characteristics of the previously used technologies may thus play an important role in the analogical perception and interpretation of new technologies. Actually, if one enlarges the perspective and considers individuals rather than users, the categories that people build may have little to do with technological considerations. For instance, when discovering a new technology, a category such as "means to explain to others the person I am" may lead the user to activate perception and interpretation mechanisms related to this functional value rather than to the technical characteristics of the technology.

The insight for supporting appropriation is that, in addition to guide design and, in particular, to think up interfaces that users will find intuitive, analogies could be used to draw users' attention to how a given technology can have similar or very different important (rather than superficial) characteristics to other technologies that they may know and use. This may be used to undermine technically misleading categories and/or categorization processes.

7.6 Addressing Explanations in AI-Based Technologies

A basic implication of the emancipation concerns I raised in chapter 1 is that as users, we have the right to both know if we use AI-based technologies and understand why they produce the results they offer us and/or the decisions they take—that is, to receive pertinent explanations. This has become a major concern. Since the development of deep-learning techniques and now generative systems based on large language models, we almost all use AI-based applications.

Being aware that we use AI-based technologies is not straightforward. Some of our uses of these technologies are conscious and voluntary. For instance, using chatbots (e.g., ChatGPT) or recognition systems (e.g., smartphone applications to identify plants or animals from photos) is a personal choice. However, we are also often unaware that we are using AI technologies, and it is indeed difficult to use the Internet without being exposed to the outputs of AI techniques. Recommendation systems have become ubiquitous, and, when online, it is practically impossible to watch a video or read a review of a book without being proposed other videos or books that may be of interest to us. Similar AI techniques decide what posts or news we are targeted with. The emails we receive have been scanned by AI-based anti-spam filters (and also other filters). In the near future, generative systems will probably become a basic if not dominant way to automatically produce answers to consumer enquiries. And while some providers use AI as an argument, this is not the case for the role AI techniques play in the news we receive or the management of our insurances or banking actions. This state of affairs is an issue, if nothing else because, as users, we sustain the values and principles these systems convey. For instance, by using news feeds and search engines including AI features, we accept and perpetuate the fact that people are provided with information that has been selected or influenced by algorithms or, rather, the information that algorithm designers or promoters think is the best for these people. Actually, we actively contribute to this process by clicking on one proposal or another, or by providing data to feed the learning mechanisms (and we do this for free, which sustains big tech companies' benefits).

In the following sections, I will focus on how the appropriation perspective developed in this book provides specific insights for addressing the *explanation* issue. For more general HCI works considering the philosophical, social, legal, and technical aspects of AI technologies fairness and transparency, see for example Abdul et al. (2018), and see Karanasios et al. (2021) for an analysis of how activity theory may help.

Usefulness and Relevance

As for other applications, the appropriation of AI technologies is driven by the attributed functional values. Identifying these values is thus core.

At the time of writing this book generative systems such as ChatGPT are gaining importance but, as they are in constant evolution, we lack sufficient empirical studies of their uses. For instance, it seems that an emerging functional value is to use these systems to produce syntheses or essays via a two-steps process: Obtain a kind of "first thoughts" draft document that aggregates the basic contents that the text should include and then use this draft as a substrate for personal reflection and writing. In other words, the system is used as if it was producing a kind of synthesis of the outputs of a coherent set of search requests on the Web. The analyses that will be conducted in coming years will clarify this picture.

In addition to being interesting as such, the empirical studies of the use of recommendation systems (an "older" technology) may help to reflect on other AI-based systems such as ChatGPT. These studies show that when we use recommendation systems, our expectation is that the system correctly infers what is of interest to us from what it knows about us and what the market offers. For instance, perceived usefulness is positively influenced by the novelty and diversity of the proposals (Pu et al., 2012).

What surfaces here is that one of the specificities of AI-based technologies such as recommendation or generative systems is *how* we consider and evaluate their outputs. When we ask the smart speaker to put the kitchen light on, the interpretation of the result is factual and unequivocal. Although the speech recognition process involves AI techniques, the output is similar in nature to that of pressing the kitchen light switch. In direct contrast, the central issue when we ask the recommendation system for advice (or use generative AI systems) is not whether the system responds or not but the very characteristics of the response.

The "focus on how people interact with tasks rather than how they interact with technologies" principle suggests that the important characteristics of the system response is *relevance* rather than intelligence. With the exception of users who question as such the intelligence of technologies, the important matter here is the task at hand (e.g., finding a deal that satisfies us) and not the characteristics of the means (if the system is "intelligent" or not). Actually, it has been shown that when people know and understand how "intelligent" systems work, they often cease to consider them intelligent.

Moreover, although people may seem to interact with these technologies in a very similar way to how they interact with humans, detailed analyses reveal substantial differences (Porcheron et al., 2018; Purington et al., 2017).

However, as we use our own (human) intelligence throughout our interactions with these technologies, the general mechanisms influencing how we receive recommendations or solutions to problems play a role. For instance, accessing e-commerce recommendation systems via smart speakers conflicts with some of the factors affecting trust, with an effect on appropriation (Rabassa et al., 2022): While it has been shown that trust is improved by the inclusion of familiar items in the recommendation set (Pu et al., 2012), vocal interaction usability issues allow proposing very few items only. AI-based technologies also lead to some anthropomorphism, with unclear implications. For instance, Purington et al. (2017) identified a link between the personification of intelligent assistants and satisfaction, which raises the question of whether satisfaction leads to personification, or if people who personify devices are more likely to be satisfied with their performance. The way we evaluate the pertinence of the outputs of these technologies is also impacted by human reasoning biases. For instance, the confirmation bias (i.e., our tendency to favor information that confirms or supports our personal beliefs and values) suggests that, unfortunately, we may be more satisfied with a news feed that confirms our views rather than one that makes us think. And, here again unfortunately, the providers of these news feeds are private companies whose best interest is to keep us satisfied.

The general takeaway is that how we evaluate the relevance of the responses of these systems stems from an interplay between how we evaluate the responses as such, with our individual perceptions and cognitive biases, and how we consider the "intelligent" nature of AI-based systems. Let's now go into more detail.

From Information to Explanation

Studies show that whether or not AI-based technologies allow users to understand the rationale for their outputs is an important factor in their appropriation (Miller, 2019). This requirement can be met by presenting how they produce these outputs (transparency). However, this does not work. The issue is not limited to the cognitive burden of the process complexity or, here again, the possible mismatch with user conceptualizations (Eiband et al., 2018). The problem is that these systems proceed very differently from our ways of doing, and when we are surprised by or doubt the output and require evidence, examining how they proceeded provides little support. In other words, presenting the system process is not an explanation. AI-based technologies must thus allow users to understand the rationale for their outputs by generating a relevant explanation, which typically comes in the form of an ad hoc message.

The understandings proposed in this book lead to the observation that any "explanatory message" that the technology presents is actually just information: It *becomes* an explanation if it is received as such by the user or, in other words, if it is given such a functional value. Therefore, here again, the analytical entry point should thus be the users' needs, and the elements that may play a role in the fact that the system output acts as an explanation. Hopefully, the reasons why people need explanations have been extensively studied by different social sciences. Miller (2019) proposes a recent AI-oriented synthesis: We search for explanations to learn and develop understandings, to predict or control what will happen and, in his words, to "reconcile the contradictions or inconsistencies between elements of our knowledge structures" (p. 15).

This understanding provides a rationale for the finding, already identified in the 90s and once again highlighted by Miller (2019), that it is much more efficient to contrast the systems' outputs with what the user may have expected, and/or potential alternatives, than to present how the system produced these outputs. Typically, for recommender systems, the explanation does not come from a description of how the system decided to recommend a given product, a process that may include dozens of odd characteristics and unintelligible weightings and probabilities, but instead results from detailing why this product was proposed *rather than* another that we know about, or others that seem similar. Rather than only knowing why the system thinks we have this specific disease, we also want to be reassured about more severe possibilities we had in mind and sometimes understand why we should unfortunately abandon our hope for a less severe illness. Answering *Why not?* and *What if?* questions rather than simply *Why?* has proven to increase the understanding and acceptation of the system outputs or, in other words, the attribution of a functional value "explanation" to the information provided by the system.

The perspective proposed in this book also provides some rationale and helps to address another point made by Miller (2019). The process by which people select information may be (in his words) "biased," and the causes retained by users as explanations differ from expectations. This may be reframed by considering that users actually address two different tasks: (1) obtain some results from the system (e.g., some recommendations), which is the motivation for using the technology; and (2) make up their minds about the results using the explanatory information (for lack of a better term) produced by the system. As we saw in chapter 3, these tasks must be dissociated and considered as precisely as possible. For instance, with respect to what the technology is used for, finding the product that best corresponds to one's needs and identifying a list of options to be regarded in more detail when going to "real" shops are two close but different tasks. What will be received as an explanation may vary from one case to another, and recommender systems should thus (also) offer explanations dedicated to making up one's mind rather than just buying. Similarly, with respect to making one's mind up, checking the fit of the recommended item with one's needs, evaluating its value for cost or wondering if one can trust the system are also different motivations/activities. According to the precise motivation(s) in question, users may focus on and attribute, for example, an "explanation" or a "reassuring" functional value to different pieces of the explanatory information produced by the system.

7.7   Complements from Other Design Perspectives

The preceding sections featured the specific insights and takeaways provided by the understanding of appropriation proposed in this book. The next section presents insights from other approaches that provide coherent pertinent complements, despite being built on very different premises in some cases. These are the technology-enhanced activity spaces, semiotic engineering, infrastructuring, automated interface adaptation, and meta-design.

Technology-Enhanced Activity Space

Acknowledging the importance of considering that (1) users act within ecologies of artifacts and (2) they should be empowered to create better environments for their professional, learning, or leisure activities, Kaptelinin and Bannon (2012) have argued for a change in the design focus. They propose that, rather than systems (i.e., products), design should focus on technology-enhanced activity spaces (TEAS)—that is, "spatially and temporally organized configuration of resources, including digital technologies, which enable an individual or a group to carry out one activity or several coordinated activities" (p. 294). The argument is very coherent with the overall perspective adopted in this book: "Designers cannot simply impose on users their understanding of how people in the setting should act and how they should use technology—even if the understanding is based on a thorough analysis of users' needs and requirements" (p. 290). When the introduction of a new application is an extrinsic practice transformation initiated by designers, Kaptelinin and Bannon advocate to allow users to accomplish their own practice transformations via integrative technologies such as "metatools" or "connectors."

Although the overall TEAS perspective is very stimulating, seeking to interconnect artifacts that were not designed for such a purpose is an open technical problem.

Nevertheless, offering straightforwardly usable adaptation means (and, when possible, some conceptual flexibility), designing new plug-ins to close conceptual or functional gaps or indexing existing plug-ins via functional values is very coherent with the important characteristics of intrinsic practice transformation raised by Kaptelinin and Bannon: continuity (i.e., users can address the original problem in the original context for an immediate transformation), directness (i.e., the artifact is straightforwardly integrated within the practice), and solutions that are situated (appropriate for the concrete task at hand and the other used technologies) rather than generic.

Moreover, and adding yet another specific interest to plug-ins, the TEAS approach may be partially addressed by designing *epiphyte meta-plug-ins*— that is, pieces of code that connect different plug-ins without modifying them. As an example, a NoteToTask meta-plug-in makes it possible to edit and attach one's thoughts via the plug-in that allows to add notes to emails and, when deciding to skip to a proper task description, to generate the task from the email and the text of the note to avoid information loss (Tchounikine, 2019a). The provision of such meta-plug-ins is a way to acknowledge how users take advantage of different notions and means (design *from* appropriation) and to open a technical substrate for creating relations between these means and their underlying notions via practices (design *for* appropriation). Use analyses at a later date may lead to implementing particular workflows (design *from* appropriation) and offer them as a new extension (new plug-ins).

Semiotic Engineering

de Souza (2005, 2014) has proposed a semiotics approach of technologies interfaces that is built on different theoretical bases than the ones I refer to in this book but intersects and complements the analyses I proposed. Semiotics is the study of signs (anything—text, image, and so on—that communicates a meaning to an interpreter) and the associated meaning-making mechanisms. The core idea of semiotic engineering is that interfaces encode a message that the designer communicates to the user: "Here is my understanding of who you are, what I've learned you want or need to do, in which preferred ways, and why. This is the system that I have therefore designed for you, and this is the way you can or should use it in order to fulfill a range of purposes that fall within this vision" (de Souza, 2014, p. 25). When developing interfaces, designers thus develop hypotheses about how users will interpret this message, yet this is sometimes limited to the implicit expectations that (1) all the users will have the same interpretation and (2) this interpretation is identical to that of the designers and/or the users involved in the usability tests. An example of semiotics analysis may be found in Chagas et al.'s (2019) study of how commercially available smart home IoT devices are appropriated. They consider that "appropriation corresponds to the stable state of interpretation which allows users to organize a rich and productive set of interactive discourses with the technology at hand" (p. 11). The semiotics perspective draws attention to how users engage in a "conversation with the technology" and answer issues such as "What's this?" "What happened?" (e.g., when one of the devices triggered an unexpected action) or indeed "Why didn't it happen?"

The semiotic approach provides an additional perspective to the inter- and intra-user variability issue I introduced in section 7.1. First, the meaning-making mechanisms of different users may differ (inter-user variability issue). Second, as meaning is an ongoing process that only stabilizes temporarily, how a given user interprets the message may also evolve (intra-user variability).

This approach also provides a conceptual tool for designing adaptation means. de Souza and Barbosa (2006) propose considering the different forms of interface adaptations that could be offered to users in terms of lexical items, grammatical structures, and meanings. Examples include changing only lexical terms, for example renaming or aliasing an icon or a menu item; changing only the grammatical structure—for example, moving from a "select a piece of text and then the action to be proceeded on the text" pattern to a "select the action and the piece of text" pattern; changing the meaning without changing the lexical terms or the grammar—for example, defining an additional action for the "print" label that allows the user to print all the files of a folder; changing both the meanings and lexical terms, but not the grammar—for example, defining a new formatting style; or changing meanings, grammatical structures, and lexical items—for example, defining a "search style" feature that, rather than returning one occurrence at a time, presents them all (p. 11).

Another pertinent insight of the semiotics perspective is the importance of maintaining coherence. Some adaptations (e.g., renaming some items of a menu and/or reorganizing the menu) may actually break down the overall consistency of the interface message and prove to be problematic.

Finally, semiotics also proposes a complementary way of considering the role of categories suggested by the analogical perspective (see section 7.5) and the way it may be used to inform training sessions or design. Taking a semiotic perspective, meaning-making activities are based on abductive reasoning. We build interpretations of signs (interpretations of the word) and act accordingly until evidence contradicts them, obliging us to revise our hypotheses. This explains why users may develop uses deriving from an understanding of the technology that is completely wrong, yet stick to their interpretation as long as it allows them to conduct their activity to their satisfaction (de Souza, 2014). Actually, as long as their hypotheses allow them to conduct their activity to their satisfaction, they have no reason to question their interpretation. And although some users may suspect that their interpretation may not be the most efficient, the cost-benefit of developing a new one (as they perceive it) may not be worth engaging in new abductive efforts or, returning to analogies, worth creating new categories. As I have already mentioned, this attitude can be destabilized via focused training sessions and/or smart design.

Infrastructuring

The entry point of the infrastructuring approach (Stevens & Pipek, 2018) is that appropriation is a social activity, which is "related to and expressed by the transformation of an activity system in reaction to the adoption of a new technology (p. 159)." As the conflicts or breakdowns that may develop in relation to a (new) technology are not detached from the underlying infrastructure, the focus should be on "the entirety of devices, tools, technologies, standards, conventions, and protocols on which the individual worker or the collective rely to carry out the tasks and achieve the goals assigned to them" (Pipek & Wulf, 2009, p. 455). The argument is that, in addition to the technological links, these different elements are interrelated through the uses that develop at both individual and organizational (e.g., work task or department) levels. The actual infrastructure thus develops from both the professional designers' and the users' activities.

A practical implication for design is the interest of articulating adaptation and collaboration means (Stevens et al., 2010); see also Draxler et al. (2012). While my suggestion to take a functional value perspective to user information (e.g., index plug-ins according to functional values) addresses this point in general, the infrastructuring approach proposes designing integrated software architectures. As an example, Ludwig et al. (2017) have studied the appropriation of 3D printers. The study revealed that although adapting and extending the hardware and/or software modules is in line with the design rationale and spirit of these technologies, they provide little help to do so. The authors recommend a native integration of communication and documentation means within the technology.

Automated Interface Adaptation

The focus on users as actors led me to present examples of interface adaptability—that is, how users can modify some aspects of the interface of the applications they use. A large set of HCI works have also studied adaptivity—that is, how applications may automatically adapt some aspects of their user interface. These works address goals such as adapting the interface to different displays (e.g., smartphones and laptops), contexts (e.g., mobility), or types of users by exploiting "user models" (which usually encompass general characteristics such as "gender," "age," "culture," or "impairments"), "environment models" (e.g., stationary or mobile context, light conditions, or noise level), and/or data acquired from user behavior; see Abrahão et al. (2021) for a synthesis.

Todi et al. (2021) highlight that "picking an adaptation can be considered a hypothesis on how useful it is for user," that "utility is . . . non-stationary," and that "adaptive systems could provide greater benefits by planning sequences of adaptations that gracefully lead a user through gradual changes" (p. 2). This brings them to formulate the problem of adaptation (e.g., the adaptation of home screens, graphical layouts, or application menus) as a stochastic sequential decision-making problem, which they address via a model-based reinforcement learning approach (a specific machine-learning technique).

Although it makes sense to consider adaptation as a decision-making problem, the analyses developed in this book suggest that it cannot be solved automatically, if only because of (1) the complexity of the factors driving user-situated needs and (2) the fact that many of these do not have any technical inscription in the system. The implication is that, rather than supporting appropriation by aligning the interface with the actual practice, automated adaptation may lead to one or several of the (numerous) issues revealed by empirical studies such as incorrect capture and/or interpretation of end users' needs or user cognitive disruption (Abrahão et al., 2021).

Paradoxically, limited and asemantic adaptations that can be controlled by the user may prove to be more useful and support appropriation better than approaches based on (often arguably naive) user models. For instance, using the menu example, user appropriation may lead to specific use patterns, which may serve as a basis to *propose* a reorganization of the menu. What is at play here is, modestly, the automated identification of the technical traces of one of the outputs of the appropriation (the use patterns).

Meta-Design

Meta-design is a radical approach to software adaptation. This engineering perspective argues that users, as the owners of problems, should be empowered to act as designers (Fischer, 2003; Fischer et al., 2004). Rather than considering users' needs and/or more or less involving them in the design, which is at the core of UCD and PD approaches, meta-design argues that users should become codesigners at the time of designing the technology and remain involved for as long as the system exists. For this purpose, the meta-design manifesto (Fischer et al., 2004) advocates the general SER model I have already mentioned: seeding (an initial professional development provides a seed and the means to complete or extend the system), evolutionary growth (users adapt the system and continue its development), and reseeding (professional development that organizes the changes developed during the evolutionary growth).

This macroscopic perspective proposes a general framework for empowering communities of users. A core principle of the meta-design approach is to create "the enabling conditions for collaborative design in which all participants, not just skilled computer professionals, incrementally acquire ownership of problems and contribute actively to their solutions" (Fischer et al., 2004, p. 36). An important success factor is a domain within which users act within a culture of participation and are provided with the means to participate actively (Fischer, 2010). Fischer et al. (2017) insist that the meta-design approach is first and foremost about sustaining a cultural transformation, and they draw links with other movements such as Democratizing Innovation, Wikinomics (mass collaboration) or the Maker Movement.

In a nutshell, the key takeaways of this chapter include:

- Current design approaches such as UCD or PD should and can be enhanced to consider appropriation concerns.
- The theoretical understandings presented in this book highlight different points of attention such as the inter- and intra-user variability issue or the role of conceptualizations and analogies.
- These understandings also provide different conceptual tools that may help to provide users with pertinent adaptation means—for example, the notions of instrumental direction or meta-plug-in.

8   Conclusion

How and why people appropriate digital technologies is a vast, complex, and multidisciplinary research field. This book has proposed a general picture and a theoretical understanding. This final chapter argues that appropriation should be considered a central notion of human–computer interaction (section 8.1), reviews some of the points made throughout the text (section 8.2) and the research directions they suggest (section 8.3), and finally comes back to emancipation considerations (section 8.4).

8.1   Appropriation Should Be Considered a Central Notion of HCI

In most of the works related to the analysis of how people act with digital technologies and, more generally, in the HCI field, if and how people appropriate technologies is addressed superficially only, when not ignored. Emphasis is generally on adoption and/or usability.

I argue for considering appropriation as a structuring element.

Although adoption is a notion that is indeed useful to the understanding of digital technologies uses, it is essentially a matter of interest for the designers and promoters of the technology, and not for people. Moreover, it suggests a frame of mind separating design and use.

Usability is also a useful notion, which draws attention to the pertinent question of whether a system can be used. However, here again, usability is first and most a matter of interest for designers. Although users do of course have an interest in being able to use a product, the main appeal is not the use of a given technology but the ability to conduct their activities as they want—that is, according to their perceptions, goals, and criteria. Technological means and their characteristics come into the picture within, and according to, this context. Moreover, while addressing usability brings analysts and designers to consider users, it remains a technocentric analytical entry point. People are people, not users. Considering them simply (or firstly) as users intrinsically puts emphasis on the individual-system interaction, which, as we saw, is an analytical perspective that presents intrinsic limits. Usability is not the driving force in how uses develop; it is just one of the factors that play a role. Moreover, usability is a multidimensional construct, which is not without problems (Tractinsky, 2018).

In direct contrast, the appropriation notion puts the emphasis on what the interest of technologies is for people, namely the way they mediate and extend their possibilities in practices. The implications go much further than the general principle stating that technologies must be considered by taking an activity-centric perspective. Focusing on appropriation calls for addressing how people use technologies as a specific aspect of a general and pervasive phenomenon that is rooted in our human nature: We develop resources for ourselves, and we develop these resources within the context of our interactions with the tasks we consider. It calls for starting from, and prioritizing, the consideration of what people do, the different contextual or general forces that apply, the involved cognitive mechanisms, and, in this context, the properties of the artifacts. These aspects must be addressed as a whole. As we saw, focusing on the characteristics of technologies or, conversely, only considering technologies at an abstract level (e.g., "social media") does not permit a detailed understanding of uses. Conversely, although the psychological mechanisms and forces at play (e.g., perception, meaning-making mechanisms or ways of doing) do not stem from technologies, how they apply when people use technologies is influenced by the characteristics of the latter. And although it is impossible to design the user's interaction with the system, which emerges in action, it is possible to act on its technical substratum. Adoption and usability issues remain in the picture but are not the driving concerns.

This general framework may be explored according to different perspectives. The actor-centered developmental perspective proposed in this book is one such approach. It does not exclude the use of others. As mentioned in chapter 1, appropriation is a complex phenomenon, and multiple points of view are required to understand complex phenomena.

Let me stress once again the importance of considering designing *for* appropriation. The evolution of the technical ecosystem toward the offer of myriad applications we can select and use tends to generate the following frame of mind: Let's leave designers free to design as they want, just taking care that their systems are usable. Users will do what they want with this, and designers will see what happens and react. This frame of mind, which may be backed by arguments such as the complexity of appropriation processes or the difficulty of understanding what causes actual uses of technology, is convenient for designers. However, without an in-depth comprehensive understanding of the phenomena at play, a "see what happens and react" process is unlikely to address appropriation issues. The observed (or log-inferred) sequences of actions processed by users, and their technical adaptations, if any, are simply projections of the outputs of users' appropriations on the technical substratum.

In some sense, to make the individuals solve alone the issue of making technology their own is to refuse to solve the problem of producing designs that empower people and respect them. Indeed, appropriation is a process that is developed by individuals. However, this development is both supported and constrained by the technological substratum, which is defined by designers. The fertility of this substratum is the designers' responsibility. Moreover, as we saw, the different forces underlying appropriation have strong, and in some cases, problematic and harmful impacts. This cannot be ignored, in particular for workers who are obliged to use certain applications and/or people who cannot escape the social pressure of, for instance, using one social network or another.

8.2 Appropriation in a Nutshell

Rather than attempting to summarize all the preceding chapters in a few lines, this section compiles a selection of points. The adopted bullet-list presentation highlights clear-cut structuring ideas, making it easier to consider and discuss them as such and to identify what additional work is required. However, this presentation also breaks complex arguments and analyses down into pieces. These points must thus be understood as presented, in context and with additional explanations and precautions, in the preceding chapters.

Definitions
1. Appropriation is the process by which users, while interacting with the tasks they consider, attribute functional values to digital artifacts and

associate them with one or more patterned way(s) of (1) addressing the task at hand and (2) using the technology; the outputs include integration in practices and functional transparency.

2. The functional value of an artifact for an actor is the utility of this artifact for achieving some tasks or goals as perceived by this actor.

3. Functional transparency is the fact that, when an individual is confronted with a task, an artifact (or set of artifacts) and the associated psychological constructions are immediately and without any conscious explicit effort mobilized.

Explanation of the Phenomenon

1. Appropriation is a developmental process. The fact that people develop is the underlying mechanism, and not a by-product, of appropriation.

2. While people interact with the tasks they consider, they create resources to address these tasks. This stems from an ontological characteristic of human activity: Activity is both productive and constructive.

3. Artifacts that are perceived as useful are attributed one or several functional value(s) and used as mediators. They become part of the user's ecology of artifacts—that is, the set of technologies through which people act.

4. The attribution of functional values is driven by people's actual activity/motives (which may correspond to or differ from expectations) and by their perception and understanding of their actual practices. It is thus shaped and/or impacted by all the different social (e.g., institutions or work practices), psychological (e.g., conceptualizations, ways of addressing tasks or meaning-making mechanisms), and technical (e.g., application features or adaptation means) forces and factors that may play a role in these perceptions and practices. Rather than considering what people build from a given technology, the focus must be on what they build from and for their activities when these activities are partially mediated by the technology in question.

5. Activity (what people do), rather than epistemic aspects (what people know), plays a driving role. The key aspect is not the users' understanding of the applications or devices they use but their perceived functions. Activity relies on knowledge and conscious processes but is not subordinated and limited to them.

6. The resources that are created or modified by users include the artifact(s) and cognitive constructions.

7. A possible model is to address these resources as quadruplets (task, way of doing, artifact, way of using)—that is, the task that is considered by the individual (or the set of tasks when the considered one interplays with others); the cognitive structures shaping the patterned way(s) of addressing these tasks; the associated artifact(s); and finally, the cognitive structures shaping the patterned way(s) of using these artifacts. The creation or evolution of such resources involves intertwined developments that are related to the task (the user's conceptualization of the task, ways of doing, perception and understanding of his/her achievement of the task); to the artifact (the user's conceptualization of the technology, ways of using, perception and understanding of his/her uses thereof and their outputs such as efficiency and artifact adaptations); and the task-artifact links (i.e., the functional values attributed by the user).

8. People may be (and are indeed often) involved in several interrelated activities at any one time, meaning that they simultaneously address different goals. If and how these activities/motives interplay may thus shape or impact the attribution of functional values and the uses of technologies.

9. Users, as humans, are engaged in activities stemming from fundamental psychological needs. These conscious or unconscious high-level motivations may drive the attribution of specific functional values to technologies and their uses. They may also impact the attribution of task-level functional values through their interplay with task-level motives. These high-level transversal forces include desires and needs such as aesthetics, the quest for competence or autonomy, the reflection of one's self-identity, the maintenance of self-esteem or the avoidance of anxiety and isolation; phenomena such as the attachment to, and/or psychological ownership of, the artifacts (devices, content) or the habits and uses that developed around the technology (emotional bonds); or voluntary engagements such as the quest for authenticity. The previous list is not exclusive, and several of these aspects interrelate in different ways.

10. One-off appropriations—that is, the contextual use of an artifact to address a task—stems from (1) an existing association of the task with a set of resources in the general sense (i.e., artifacts and ways

of doing/using) and (2) given the unavailability of the artifact, the search for new contextually contingent functional links. This search is shaped by the usual resource characteristics and particularly the associated way of doing. Such new functional links may be temporary or become part of the stable yet evolutive configuration of internal and external resources for the task at hand.

Criteria and Symptoms

1. Integration in practices is a symptom rather than a criterion of appropriation. Its intrinsic limitation as a criterion is that it denotes a characteristic of use that stems from appropriation rather than the process as such, and thus it does not discard other phenomena (such as constrained uses) that produce similar outputs.

2. Functional transparency is a more fundamental, precise, and measurable criterion of appropriation than integration in practices. An application or device can be said to be appropriated when the user mobilizes it immediately and without any conscious explicit effort when he/she considers a given task. This criterion is met when the user has attributed one or several functional values to the application or device and associated it with imported, adapted, or developed-in-this-context psychological constructions, namely ways of addressing the task and ways of using the technology. In other words, the stable-for-now technical + psychological resource has become a (or part of a) functional organ for the user.

3. Use transparency—that is, the fact that using the system does not generate an interfering conscious process—is a likely symptom of appropriation but a contingent aspect thereof. Appropriation usually—though not necessarily—leads to frequent uses, with an effect on use transparency.

4. Artifacts adaptation is a possible symptom of appropriation but a contingent aspect thereof. What is intrinsic to appropriation is user evolution. Technical adaptations, if any, occur in relation to the user's ways of doing/using and are driven by them. Nevertheless, adaptations addressing the improvement of the activity–technology fit are both possible symptoms of, and means for, appropriation.

5. Efficiency is a contingent aspect of appropriation.

Implications for Use Analyses

1. People do not use the technology but rather their appropriations thereof.

2. As uses stem from what has been developed through appropriation (attribution of functional values, functional transparency, ways of doing/using), understanding uses requires first and foremost the identification and analysis of the user's actual activities and motives.

3. The focus must be on how users interact with the tasks they consider rather than on how they interact with technologies.

4. Understanding uses requires considering precise functional values and being attentive to the fact that the artifact-functional value relation is not necessarily a 1:1 relation.

5. Appropriation phenomena are intrinsic to use and thus apply in all cases, regardless of whether the design is "good" or "bad." The use of technologies for unexpected tasks is just one case of the general and pervasive appropriation phenomenon.

6. Appropriation phenomena are not limited to newly introduced technologies. The technologies that an individual uses may be given new functional values or stripped of previous functional values. Moreover, while the functional value remains stable, the psychological constructions shaping the way of using may evolve. New uses, but also new user-artifact tensions, may thus occur.

7. Appropriation involves intertwined factors and forces that differ in nature. The use of just one analytical perspective is unlikely to allow a proper understanding.

Implications for Design

1. Design for appropriation is to be thought of as efforts to increase the likelihood that appropriation will develop by facilitating, or at the very least not hindering, the underlying psychological and technical processes.

2. Design must address inter-user and intra-user variability.

3. Inter-user and intra-user variability can only be addressed by offering adaptation means that allow users to make the technology (the media) congruent with their conceptualizations, ways of doing, needs, and/or concerns.

4. Several standard adaptation techniques (e.g., renaming items, customizing interfaces, and defining specific behaviors) are good steps in this direction.

5. The design of adaptation techniques may be improved by being attentive to users' ZPD (e.g., by helping users to close the gap between the notions underlying present practices and those underlying the proposed means), offering conceptual variability and flexibility, favoring productive analogies, offering under-designed means to engage users in adaptations, and using functional values and instrumental directions to (1) draw users' attention to potentially useful means and (2) support the sharing of experiences.

6. The plug-in technique presents specific intrinsic advantages. These include extending, rather than changing, the existing technology; allowing self-initiated and evolutionary (rather than revolutionary) changes; avoiding design dilemma and, for instance, offering several alternative plug-ins that involve different notions and/or different technical skills, thus increasing the chances that some of these adaptation means will be available in the space within which the user may act productively.

8.3   Research Directions

The above synthesis suggests the need to further explore several research directions, and the first one may be the articulation of individual and collective dimensions. In this book, I took individuals as the analytical entry point. Collective dimensions are acknowledged and considered from this perspective. What emerges from the previous bullet-list presentation is that these dimensions are present practically everywhere (e.g., socially carried ways of doing, work practices, or role of peers in developmental aspects and technical adaptations). This does not downplay the specific interest and, in my opinion, the necessity of considering individuals. Rather, it confirms that *complementary* understandings may be gained by analyses that take collectives as analytical entry points or, at the least, consider them as such. In this respect, I argue for the importance of the development of social perspectives that do acknowledge the points made in this book and the fact that people are above all individuals, with their personal motives, needs or perceptions.

Although the collective dimension of activity already underlies the works by Vygotsky and Leont'ev I built on, an important evolution is how the third generation of activity theory (Engeström) focuses on the transformation of institutions. This has become the main analytical approach, which does indeed help to make sense of collective aspects (but, as I argued above, may lead to downplaying individual dimensions). Spinuzzi (2020) argues that third-generation AT was adapted to settings where roles, objects, and organizational boundaries were well established, and a fourth generation may be required to address the fact that, due to different factors (from which the evolution of the technical ecosystem), boundaries and structures fade away and stakeholders are less stable. According to Engeström and Sannino (2021), the challenge of a fourth-generation AT is rather to address the complexity and dynamics of interconnected objects such as poverty, climate change, and pandemics, which requires considering "coalescing cycles of expansive learning in a heterogenous coalition of activities facing a critical societal challenge" as units of analyses (p. 20). See also Karanasios et al. (2021). If these works lead to an improved general understanding of human activities, this will in turn inform the study of digital technologies appropriation.

Moreover, the centrality of the cognitive dimension of appropriation put forward in this book argues for the further exploration of (at least) two aspects. First, there is a clear need to study the cognitive mechanisms at play in more detail. This includes designing protocols for measuring functional transparency and, possibly, degrees of appropriation (i.e., the extent to which a user appropriated an application), the relevance of the scheme perspective, or the potential role of analogies. Second, close attention should be paid to the social or human needs that may play a role. This includes identifying the forces that may be involved (I featured the role of XXP, psychological ownership, or engagement to pave the way for future works), their impact on the cognitive mechanisms underlying appropriation, or their conscious/unconscious nature.

Finally, in addition to general considerations, the implications for design listed above include precise proposals—for example, using functional values and instrumental directions to index the plug-in offer, design plug-ins that offer some conceptual flexibility, link plug-ins via meta-plug-ins, or use under-design to engage users in adaptations. The next step is to further explore them via ad hoc implementations.

8.4 Supporting Emancipation

I stated in the introduction of this book that understanding appropriation and informing design accordingly was important to enhance user control and agency as well as support emancipation. I would like to come back to this point in this final section.

Technologies, as media, extend our human possibilities. The other side of the coin is that using technologies makes us adapt to them in the very literal sense (via psychological developments), impacts our ways of doing, more or less transparently nudges us, and may come into conflict with our psychological needs or engagements.

When these issues may be addressed in terms of values, I argue for addressing them in terms of emancipation. The way technologies affect us is not intrinsically good or bad. What is core is that, as individuals, we have the right to be aware of the phenomena at play and be equipped with means for understanding them. Such an understanding contributes to our control and our agency, and is thus emancipatory. Of course, as individuals, we are unlikely to identify, decipher, and make perfect sense of all the forces and factors influencing our appropriations and uses of technologies. This does not negate the positive and useful nature of any gain.

This analytical perspective makes it clear that some aspects of the technical ecosystem evolution raise important issues.

One of these issues is the way some applications literally confiscate our activities and data. Swapping from one application to another one offering the same services under different conditions should be granted and basic. As pointed out by Yalom (1980), part of the difficulty of taking decisions is assuming responsibility for not having decided earlier. Although this point addresses the big decisions we take (or do not take) in our lives, it may be transposed to the issue we consider here. When using an application conflicts with activities, needs, personal values, or engagements, the lack of import/ export features should not be a factual or a (possibly bad faith) justifiable reason for continuing to use this application. More generally, designs and/or technical implementations unnecessarily constraining users to stick to one technology should be avoided.

The current state of affairs unfortunately has some problematic similarities with the "invisible hand" myth in economy. Most of the tasks for which we use digital technologies (e.g., managing personal information or communicating with others) can be addressed via different applications. However, although this seems to enhance our chances to access appropriate technologies that present few if any obstacles to our functional and human needs, this is not necessarily the case. The analyses of social media developed throughout the chapters illustrate this point well. As a matter of fact, these applications have turned into major communication means, with some people and, more importantly, some institutions, communicating solely via these channels. As put forward by the heteromation perspective, most of the intrinsic value is constructed by the users. Using these technologies is, for some people, a way to fulfill needs based on who they are, and is thus of core importance. Last but not least, as the extended self that develops is in some sense co-constructed with others via "likes" or other comments (Belk, 2016) it is not controlled by the individual alone, which makes how data is accessible and manageable a particularly sensitive and important issue. These are all major arguments supporting the point that access and interoperation should be granted and easy. Unfortunately, as we have seen, this is not what happens: applications are designed to create specific channels, and interactions are only accessible via one particular application. It is far from being technically impossible to solve this interoperability issue. It is just not in the best interest of the tech companies. The "if you don't like the application, don't use it" stance is a major issue, in particular when one realizes that, when using these technologies, one unwittingly sustains the structures underlying their rationale, production, and management.

Similar concerns apply to how applications or services are automatically customized—that is, when the system dynamically adjusts the system interface or behavior in a way that is intended to support the user. And this is particularly true when the adjustment is based on AI techniques. As we saw, search engines exploiting learning mechanisms provide an individualized service, but we have no control of this individualization and are provided with few explanations. The search engine is less our own than that of the company kindly offering it to us, and the interests at stake are not only ours.

We, as individuals, have the right to understand why and how our uses develop. We have the right to be given the opportunity to benefit from the positive outputs of appropriation such as functional transparency (i.e., transparent access to means that we perceive as useful and adapted) and its correlates, for example use transparency (we can focus on the task rather than on the means). And we have the right to use technologies that we can adjust to our evolving needs and our values, rather than having to adapt ourselves and/or our activities to the technology. The understanding of appropriation and the design insights proposed in this book are a contribution to these objectives. And a further way to make the technical offer evolve toward user emancipation is helping the aforementioned invisible hand with some social pressure. Actions may vary from asking for micro-adjustments (e.g., more technical flexibility) to, more radically, acting toward empowering users and, more generally, taking care of social and economic justice.

As a final word, let's come back to the call of extending the activity-centric and functional perspectives to aspects such as psychological needs, existential concerns, or engagements. The forces stemming from what we are as humans are powerful and should be considered with great precaution, and great care. As we saw, there are some users who act with technologies within a general conscious form of engagement. HCI research should take these "existential users" into account and, more generally, should take into account all those who endorse different philosophical perspectives. This, however, is not sufficient. Designers must not evade their responsibility to consider the needs of all the users who are likely to face HCI issues related to fundamental human needs or existential questions—which, as we saw, may well be all of us.

References

Abdul, A., Vermeulen, J., Wang, D., Lim, B. Y., & Kankanhalli, M. (2018). Trends and trajectories for explainable, accountable and intelligible systems: An HCI research agenda. Proceedings of the SIGCHI conference on Human Factors in Computing Systems, Article 582 (1–18).

Abrahão, S., Insfran, E., Sluÿters, A., & Vanderdonckt, J. (2021). Model-based intelligent user interface adaptation: Challenges and future directions. Software and Systems Modeling, 20(5), 1335–1349.

Alter, S. (2014). Theory of workarounds. Communications of the Association for Information Systems, 34, 10–66.

Ambe, A. H., Brereton, M., Soro, A., & Roe, P. (2017). Technology individuation: The foibles of augmented everyday objects. Proceedings of the SIGCHI Conference on Human Factors in Computing Systems, 6632–6644.

Askehave, I., & Nielsen, A. E. (2006). Digital genres: A challenge to traditional genre theory. Information Technology & People, 18 (2), 120–141.

Asterhan, C. S., & Rosenberg, H. (2015). The promise, reality and dilemmas of secondary school teacher-student interactions in Facebook: The teacher perspective. Computers & Education, 85, 134–148.

Bærentsen, K. B., & Trettvik, J. (2002). An activity theory approach to affordance. Proceedings of the Nordic Conference on Human-Computer Interaction, 51–60.

Bagozzi, R. P. (2007). The legacy of the technology acceptance model and a proposal for a paradigm shift. Journal of the Association for Information Systems, 8(4), 244–254.

Bailey, D. E., & Leonardi, P. M. (2015). Technology choices: Why occupations differ in their embrace of new technology. MIT Press, Boston.

Bakhtin, M. M. (1981). The dialogic imagination: Four essays. University of Texas Press, Austin.

Bannon, L. (1995). From human factors to human actors: The role of psychology and human-computer interaction studies in system design. In R. M. Baecker, J. Grudin, W. A. S. Buxton, & S. Greenberg (Eds.), Human-computer interaction, 205–214. Morgan Kaufmann, San Francisco.

Bannon, L., Bardzell, J., & Bødker, S. (2018). Reimagining participatory design. Interactions, 26(1), 26–32.

Bardram, J. E. (1997). Plans as situated action: An activity theory approach to workflow systems. Proceedings of the European Conference on Computer Supported Cooperative Work, 17–32.

Bardram, J. E., & Bertelsen, O. W. (1995). Supporting the development of transparent interaction. Proceedings of the East-West Conference on Human-Computer Interaction, 79–90.

Barley, S. R. (1988). Technology, power, and the social organization of work: Towards a pragmatic theory of skilling and deskilling. Research in the Sociology of Organizations, 6, 33–80.

Barley, S. R., Meyerson, D. E., & Grodal, S. (2011). E-mail as a source and symbol of stress. Organization Science, 22 (4), 887–906.

Barricelli, B. R., Cassano, F., Fogli, D., & Piccinno, A. (2019). End-user development, end-user programming and end-user software engineering: A systematic mapping study. Journal of Systems and Software, 149, 101–137.

Baumer, E.P.S., Adams, P., Khovanskaya, V. D., Liao, T. C., Smith, M. E., Sosik, V. S., & Williams, K. (2013). Limiting, leaving, and (re)lapsing: An exploration of Facebook non-use practices and experiences. Proceedings of the SIGCHI Conference on Human Factors in Computing Systems, 3257–3266.

Bazerman, C. (2013). A theory of literate action: Literate action. Parlor press, Anderson.

Bazerman, C. (1994). Systems of genres and the enactment of social intentions. In A. Freedman & P. Medway (Eds.), Genre and the new rhetoric, 79–88. Taylor & Francis, London.

Beaudouin-Lafon, M. (2004). Designing interaction, not interfaces. Proceedings of the Working Conference on Advanced Visual Interfaces, 15–22.

Béguin, P. (2007). In search of a unit of analysis for designing instruments. Artifact, 1(1), 12–16.

Belk, R. W. (2016). Extended self and the digital world. Current Opinion in Psychology, 10, 50–54.

Belk, R. W. (2013). Extended self in a digital world. Journal of Consumer Research, 40(3), 477–500.

Belk, R. W. (1988). Possessions and the extended self. Journal of Consumer Research, 15(2), 139–168.

Bellotti, V., Ducheneaut, N., Howard, M., Smith, I., & Grinter, R. E. (2005). Quality versus quantity: E-mail-centric task management and its relation with overload. Human-Computer Interaction, 20 (1–2), 89–138.

Benamar, L., Balagué, C., & Zhong, Z. (2020). Internet of things devices appropriation process: The dynamic interactions value appropriation (DIVA) framework. Technovation, 89, 102082.

Beneteau, E., Boone, A., Wu, Y., Kientz, J. A., Yip, J., & Hiniker, A. (2020). Parenting with Alexa: Exploring the introduction of smart speakers on family dynamics. Proceedings of the SIGCHI Conference on Human Factors in Computing Systems, 1–13.

Bentley, R. & Dourish, P. (1995). Medium versus mechanism: Supporting collaboration through customisation. Proceedings of the European Conference on Computer-Supported Cooperative Work, 133–148.

Binder, J., Howes, A., & Sutcliffe, A. (2009). The problem of conflicting social spheres: Effects of network structure on experienced tension in social network sites. Proceedings of the SIGCHI Conference on Human Factors in Computing Systems, 965–974.

Birnholtz, J., & Ibara, S. (2012). Tracking changes in collaborative writing: Edits, visibility and group maintenance. Proceedings of the ACM Conference on Computer Supported Cooperative Work, 809–818.

Birnholtz, J., Steinhardt, S., & Pavese, A. (2013). Write here, write now!: An experimental study of group maintenance in collaborative writing. Proceedings of the SIGCHI Conference on Human Factors in Computing Systems, 961–970.

Błachnio, A., Przepiorka, A., & Pantic, I. (2016). Association between Facebook addiction, self-esteem and life satisfaction: A cross-sectional study. Computers in Human Behavior, 55(B), 701–705.

Blom, J. O., & Monk, A. F. (2003). Theory of personalization of appearance: Why users personalize their PCs and mobile phones. Human-Computer Interaction, 18(3), 193–228.

Boczkowski, P. J., Matassi, M., & Mitchelstein, E. (2018). How young users deal with multiple platforms: The role of meaning-making in social media repertoires. Journal of Computer-Mediated Communication, 23(5), 245–259.

Bødker, S. (1991). Through the interface: A human activity approach to user interface design. Lawrence Erlbaum Associates, Hillsdale.

Bødker, S., & Christiansen, E. (2012). Poetry in motion: Appropriation of the world of apps. Proceedings of the European Conference on Cognitive Ergonomics, 78–85.

Bødker, S., & Klokmose, C. N. (2012). Dynamics in artifact ecologies. Proceedings of the Nordic Conference on Human-Computer Interaction, 448–457.

Bødker, S. & Klokmose, C. N. (2011). The human-artifact model: An activity theoretical approach to artifact ecologies. Human-Computer Interaction, 26(4), 315–371.

Bota, H., Bennett, P. N., Awadallah, A. H., & Dumais, S. T. (2017). Self-Es: The role of emails-to-self in personal information management. Proceedings of the Conference on Human Information, Interaction and Retrieval, 205–214.

Brewer, R. N., Morris, M. R., & Lindley, S. E. (2017). How to remember what to remember: Exploring possibilities for digital reminder systems. Proceedings of the ACM on Interactive, Mobile, Wearable and Ubiquitous Technologies, 1(3), 1–20.

Camus, A. (1947). La Peste. Gallimard, Paris. (Translation: The Plague [1960]. Penguin, Harmondsworth).

Caraban, A., Karapanos, E., Gonçalves, D., & Campos, P. (2019). 23 ways to nudge: A review of technology-mediated nudging in human-computer interaction. Proceedings of the SIGCHI Conference on Human Factors in Computing Systems, Article 503 (1–15).

Carroll, J. (2004). Completing design in use: Closing the appropriation cycle. Proceedings of the European Conference on Information Systems, 337–347.

Carroll, J., Howard, S., Peck, J., & Murphy, J. (2002). A field study of perceptions and use of mobile telephones by 16 to 22 year olds. Journal of Information Technology Theory and Application, 4(2), Article 6.

Carroll, J., Howard, S., Vetere, F., Peck, J., & Murphy, J. (2001) Identity, power and fragmentation in cyberspace: Technology appropriation by young people. Proceedings of the Australian Conference on Information Systems, Paper 6.

Carroll, J. M. (2014). Human computer interaction—brief intro. The Encyclopaedia of Human-Computer Interaction, Chapter 2. https://www.interaction-design.org /literature/book/the-encyclopedia-of-human-computer-interaction-2nd-ed/human -computer-interaction-brief-intro.

Carroll, J. M., Kellogg, W. A., & Rosson, M. B. (1991). The task-artifact cycle. In J. M. Carroll (Ed.), Designing interaction: Psychology at the human-computer interface, 74–102. Cambridge University Press, New York.

Carter, M., Petter, S., Grover, V., & Thatcher, J. B. (2020). IT identity: A measure and empirical investigation of its utility to IS research. Journal of the Association for Information Systems, 21(5), 1313–1342.

Cecchinato, M. E., Cox, A. L., & Bird, J. (2015). Working 9–5? Professional differences in email and boundary management practices. Proceedings of the SIGCHI Conference on Human Factors in Computing Systems, 3989–3998.

Cecchinato, M. E., Sellen, A., Shokouhi, M., & Smyth, G. (2016). Finding email in a multi-account, multi-device world. Proceedings of the SIGCHI Conference on Human Factors in Computing Systems, 1200–1210.

Chagas, B. A., Redmiles, D. F., & de Souza, C. S. (2019). Signs of appropriation: A semiotic account of breakdowns with IoT technology. Journal on Interactive Systems, 10(2), 3–19.

Choi, N., Chengalur-Smith, I., & Nevo, S. (2015). Loyalty, ideology, and identification: An empirical study of the attitudes and behaviors of passive users of open source software. Journal of the Association for Information Systems, 16(8), 674–706.

Clemmensen, T., Kaptelinin, V., & Nardi, B. (2016). Making HCI theory work: An analysis of the use of activity theory in HCI research. Behaviour & Information Technology, 35(8), 608–627.

Clot, Y. (2009). Clinic of activity: The dialogue as instrument. In A. Sannino, H. Daniels & K. Gutiérrez (Eds.), Learning and expanding with activity theory, 286–302. Cambridge University Press, Cambridge.

Clot, Y., & Faïta, D. (2000). Genres et styles en analyse du travail: Concepts et méthodes. Travailler, 4, 7–42 (in French).

Coe, R. M. (1994). "An arousing and fulfillment of desires": The rhetoric of genre in the process era—and beyond. In A. Freedman & P. Medway (Eds.), Genre and the new rhetoric, 181–190. Taylor & Francis, London.

Costabile, M. F., Mussio, P., Parasiliti Provenza, L., & Piccinno, A. (2008). End users as unwitting software developers. Proceedings of the International Workshop on End-user Software Engineering, 6–10.

Craig, K., Thatcher, J. B., & Grover, V. (2019). The IT identity threat: A conceptual definition and operational measure. Journal of Management Information Systems, 36(1), 259–288.

Crowell, S. (2017). Existentialism. In E. N. Zalta (Ed.), The Stanford encyclopedia of philosophy. https://plato.stanford.edu/archives/win2017/entries/existentialism.

Csikszentmihalyi, M. (1993). Why we need things. In S. Lubar & W. D. Kingery (Eds.), History from things, 20–29. Smithsonian Institution Press, London.

Dabbish, L. A., & Kraut, R. E. (2006). Email overload at work: An analysis of factors associated with email strain. Proceedings of the ACM Conference on Computer Supported Cooperative Work, 431–440.

Daniellou, F., & Rabardel, P. (2005). Activity-oriented approaches to ergonomics: Some traditions and communities. Theoretical Issues in Ergonomics Science, 6(5), 353–357.

Davis, F. D. (1989). Perceived usefulness, perceived ease of use, and user acceptance of information technology. MIS Quarterly, 13(3), 319–340.

de Certeau, M. (1980). L'invention du quotidien, tome 1: Arts de faire, tome 2: Habiter, cuisiner. Gallimard. (Translation: The practice of everyday life [2011]. University of California Press, Oakland.)

Deci, E. L., & Ryan, R. M. (2000). The "what" and "why" of goal pursuits: Human needs and the self-determination of behavior. Psychological Inquiry, 11(4), 227–268.

Denegri-Knott, J., Watkins, R., & Wood, J. (2012). Transforming digital virtual goods into meaningful possessions. In M. Molesworth & J. Denegri-Knott (Eds.), Digital virtual consumption, 76–91. Routledge, New York.

de Sanctis, G., & Poole, M. S. (1994). Capturing the complexity of advance technology use: Adaptive structuration theory. Organization Science, 5(2), 121–147.

de Sanctis, G., Poole, M. S., Zigurs, I., DeSharnais, G., D'Onofrio, M., Gallupe, B., Holmes, M., et al. (2008). The Minnesota GDSS research project: Group support systems, group processes, and outcomes. Journal of the Association for Information Systems, 9 (10–11), 551–608.

de Souza, C. S. (2014). Semiotics. The encyclopedia of human-computer interaction, Chapter 25. https://www.interaction-design.org/literature/book/the-encyclopedia-of -human-computer-interaction-2nd-ed/semiotics.

de Souza, C. S. (2005). The semiotic engineering of human-computer interaction. MIT Press, Cambridge.

de Souza, C. S., & Barbosa, S. (2006). A semiotic framing for end-user development. In H. Lieberman, F. Paternò, & V. Wulf (Eds.), End user development, 401–426. Springer, Dordrecht.

de Vito, M. A., Birnholtz, J., & Hancock, J. T. (2017). Platforms, people, and perception: Using affordances to understand self-presentation on social media. Proceedings of the ACM Conference on Computer Supported Cooperative Work and Social Computing, 740–754.

Dix, A. (2007). Designing for appropriation. Proceedings of the British HCI Group Conference, Volume 2, 28–30.

Dourish, P. (2017). The stuff of bits: An essay on the materialities of information. MIT Press, Cambridge.

Dourish, P. (2006). Re-space-ing place: "Place" and "space" ten years on. Proceedings of the Conference on Computer Supported Cooperative Work, 299–308.

Dourish, P. (2003). The appropriation of interactive technologies: Some lessons from placeless documents. Computer Supported Cooperative Work, 12, 465–490.

Draxler, S., Stevens, G., Stein, M., Boden, A., & Randall, D. (2012). Supporting the social context of technology appropriation: On a synthesis of sharing tools and tool knowledge. Proceedings of the SIGCHI Conference on Human Factors in Computing Systems, 2835–2844.

Ducheneaut, N., & Bellotti, V. (2001). Email as habitat: An exploration of embedded personal information management. Interactions, 8(5), 30–38.

Eiband, M., Schneider, H., Bilandzic, M., Fazekas-Con, J., Haug, M., & Hussmann, H. (2018). Bringing transparency design into practice. Proceedings of the International Conference on Intelligent User Interfaces, 211–223.

Ekbia, H. R., & Nardi, B. (2017). Heteromation, and other stories of computing and capitalism. MIT Press, Cambridge.

Ems, L. (2014). Amish workarounds: Toward a dynamic, contextualized view of technology use. Journal of Amish and Plain Anabaptist Studies, 2(1), 42–58.

Engeström, Y. (2009). The future of activity theory: A rough draft. In A. Sannino, H. Daniels & K. D. Gutiérrez (Eds.), Learning and expanding with activity theory, 303–328. Cambridge University Press, New York.

Engeström, Y. (2001). Expansive learning at work: Toward an activity theoretical reconceptualization. Journal of Education and Work, 14(1), 133–156.

Engeström, Y. (1987). Learning by expanding: an activity-theoretical approach to developmental research. Orienta-Konsultit, Helsinki.

Engeström, Y., & Sannino, A. (2021). From mediated actions to heterogenous coalitions: Four generations of activity-theoretical studies of work and learning. Mind, Culture, and Activity, 28(1), 4–23.

Evans, D. (2009). Appropriation. MIT Press, Cambridge.

Finneran, C. M., & Zhang, P. (2003). A person-artefact-task (PAT) model of flow antecedents in computer-mediated environments. International Journal of Human-Computer Studies, 59(4), 475–496.

Fischer, F., Kollar, I., Stegmann, K., & Wecker, C. (2013). Toward a script theory of guidance in computer-supported collaborative learning. Educational Psychologist, 48(1), 55–66.

Fischer, G. (2010). End-user development and meta-design: Foundations for cultures of participation. Journal of Organizational and End User Computing, 22(1), 52–82.

Fischer, G. (2003). Meta-design: Beyond user-centered and participatory design. Proceedings of HCI International, 88–92.

Fischer, G., Fogli, D., & Piccinno, A. (2017). Revisiting and broadening the meta-design framework for end-user development. In F. Paternò & V. Wulf (Eds.), New perspectives in end user development, 61–97. Springer, Cham.

Fischer, G., Giaccardi, E., Ye, Y., Sutcliffe, A. G., & Mehandjiev, N. (2004). Meta-design: A manifesto for end-user development. Communications of the ACM, 47(9), 33–37.

Folcher, V. (2003). Appropriating artifacts as instruments: When design-for-use meets design-in-use. Interacting with Computers, 15(5), 647–663.

Friedman, B., & Hendry, D. G. (2019). Value sensitive design: Shaping technology with moral imagination. MIT Press, Cambridge.

Gallagher, S. (2014). Phenomenology. The Encyclopaedia of Human-Computer Interaction, Chapter 28. https://www.interaction-design.org/literature/book/the-encyclopedia-of-human-computer-interaction-2nd-ed/phenomenology.

Gaver, W. (1991). Technology affordances. Proceedings of the SIGCHI Conference on Human Factors in Computing Systems, 79–84.

Gibson, J. J. (1979). The ecological approach to visual perception. Houghton Mifflin, Boston.

Giddens A. (1984). The constitution of society: Outline of the theory of structure. University of California Press, Berkeley.

Grevet, C., Choi, D., Kumar, D., & Gilbert, E. (2014). Overload is overloaded: Email in the age of Gmail. Proceedings of the SIGCHI Conference on Human Factors in Computing Systems, 793–802.

Griggio, C. F., Mcgrenere, J., & Mackay, W. E. (2019). Customizations and expression breakdowns in ecosystems of communication apps. Proceedings of the ACM on Human-Computer Interaction, Article 26 (1–26).

Hanrahan, B. V., Pérez-Quiñones, M. A., & Martin, D. (2016). Attending to email. Interacting with Computers, 28(3), 253–272.

Haraty, M., McGrenere, J., & Tang, C. (2016). How personal task management differs across individuals. International Journal of Human-Computer Studies, 88, 13–37.

Harrison, S., & Dourish, P. (1996). Re-place-ing space: The roles of space and place in collaborative systems. Proceedings of the ACM Conference on Computer-Supported Cooperative Work, 67–76.

Hartson, H. R. (2003). Cognitive, physical, sensory, and functional affordances in interaction design. Behaviour & Information Technology, 22(5), 315–338.

Heitmayer, M., & Lahlou, S. (2021). Why are smartphones disruptive? An empirical study of smartphone use in real-life contexts. Computers in Human Behavior, 116, 106637.

Hirsch, P. M., & Levin, D. Z. (1999). Umbrella advocates versus validity police: A life-cycle model. Organization Science, 10(2), 199–212.

Hofstadter, D. R., & Sander, E. (2013). Surfaces and essences: Analogy as the fuel and fire of thinking. Basic Books, New York.

Holtzblatt, K., & Beyer, H. (2014). Contextual design. The Encyclopaedia of Human-Computer interaction, chapter 8. https://www.interaction-design.org/literature/book/the-encyclopedia-of-human-computer-interaction-2nd-ed/contextual-design.

IFTTT (2020). https://ifttt.com.

John, B. E., & Kieras, D. E. (1996). The GOMS family of user interface analysis techniques: Comparison and contrast. ACM Transactions on Computer-Human Interaction, 3(4), 320–351.

Johnson, R., Kovács, B., & Vicsek, A. (2012). A comparison of email networks and off-line social networks: A study of a medium-sized bank. Social Networks, 34(4), 462–469.

Joinson, A. N. (2004). Self-esteem, interpersonal risk, and preference for e-mail to face-to-face communication. Cyberpsychology & Behavior, 7(4), 472–478.

Jung, H., Stolterman, E., Ryan, W., Thompson, T., & Siegel, M. (2008). Toward a framework for ecologies of artifacts: How are digital artifacts interconnected within a personal life? Proceedings of the Nordic Conference on Human-Computer Interaction, 201–210.

Kaptelinin, V. (2018). Technology and the givens of existence: Toward an existential inquiry framework in HCI research. Proceedings of the SIGCHI Conference on Human Factors in Computing Systems, Article 270 (1–14).

Kaptelinin, V. (2016). Making the case for an existential perspective in HCI research on mortality and death. Proceedings of the SIGCHI Conference on Human Factors in Computing Systems Extended Abstracts, 352–364.

Kaptelinin, V. (2014a). Affordances. The Encyclopedia of Human-Computer Interaction, chapter 25. https://www.interaction-design.org/literature/book/the-encyclopedia-of-human-computer-interaction-2nd-ed/affordances.

Kaptelinin, V. (2014b). Activity theory. The Encyclopedia of Human-Computer Interaction, Chapter 16. https://www.interaction-design.org/literature/book/the-encyclopedia-of-human-computer-interaction-2nd-ed/activity-theory.

Kaptelinin, V. (1996). Computer-mediated activity: Functional organs in social and developmental contexts. In B. Nardi (Ed.), Context and consciousness: Activity theory and human-computer interaction, 45–68. MIT Press, Cambridge.

Kaptelinin, V., & Bannon, L. (2012). Interaction design beyond the product: Creating technology-enhanced activity spaces. Human-Computer Interaction, 27(3), 277–309.

Kaptelinin, V., & Nardi, B. (2012). Affordances in HCI: Toward a mediated action perspective. Proceedings of the SIGCHI Conference on Human Factors in Computing Systems, 967–976.

Kaptelinin, V., & Nardi, B. (2006). Acting with technology: Activity theory and interaction design. MIT Press, Cambridge.

Karahanna, E., Straub, D. W., & Chervany, N. L. (1999). Information technology adoption across time: A cross-sectional comparison of pre-adoption and post-adoption beliefs. MIS Quarterly, 23(2), 183–213.

Karahanna, E., Xu, S. X., Xu, Y., & Zhang, N. A. (2018). The needs-affordances-features perspective for the use of social media. MIS Quarterly, 42(3), 737–756.

Karanasios, S., Nardi, B., Spinuzzi, C., & Malaurent, J. (2021). Moving forward with activity theory in a digital world. Mind, Culture, and Activity, 28(3), 234–253.

Keinonen, T. (2008). User-centered design and fundamental need. Proceedings of the Nordic Conference on Human-Computer Interaction, 211–219.

Kirk, C. P., & Swain, S. D. (2018). Consumer psychological ownership of digital technology. In J. Peck & S. Shu (Eds.), Psychological ownership and consumer behavior, 69–90. Springer, Cham.

Kirk, C. P., Swain, S. D., & Gaskin, J. E. (2015). I'm proud of it: Consumer technology appropriation and psychological ownership. Journal of Marketing Theory and Practice, 23(2), 166–184.

Kirk, D. S., & Sellen, A. (2010). On human remains: Values and practice in the home archiving of cherished objects. ACM Transactions on Computer-Human Interaction, 17(3), 1–43.

Ko, A. J., Abraham, R., Beckwith, L., Blackwell, A., Burnett, M., Erwig, M., Scaffidi, C., Lawrance, J., Lieberman, H., Myers, B., Rosson, M. B., Rothermel, G., Shaw, M., & Wiedenbeck, S. (2011). The state of the art in end-user software engineering. ACM Computing Surveys, 43(3), 1–44.

Koole, S. L., Greenberg, J., & Pyszczynski, T. (2006). Introducing science to the psychology of the soul: Experimental existential psychology. Current Directions in Psychological Science, 15(5), 212–216.

Lambton-Howard, D., Olivier, P., Vlachokyriakos, V., Celina, H., & Kharrufa, A. (2020). Unplatformed design: A model for appropriating social media technologies for coordinated participation. Proceedings of the SIGCHI Conference on Human Factors in Computing Systems, 1–13.

Latour, B. (2005). Reassembling the social: An introduction to actor-network-theory. Oxford University Press, Oxford.

Lau, J., Zimmerman, B., & Schaub, F. (2018). Alexa, are you listening? Privacy perceptions, concerns and privacy-seeking behaviors with smart speakers. Proceedings of the ACM on Human-Computer Interaction, 1–31.

Leont'ev, A. N. (1981). Problems of the development of mind. Progress Publishers, Moscow.

Leont'ev, A. N. (1978). Activity, consciousness, and personality. Prentice Hall, Engelwood Cliffs.

LibreOffice (2024). https://www.libreoffice.org.

Lieberman, H., Paternò, F., Kalnn M. & Wulf, V. (2006). End user development: An emerging paradigm. In H. Lieberman, F., Paternò & V. Wulf (Eds.), End user development, 1–8. Springer, Dordrecht.

Ludwig, T., Boden, A., & Pipek, V. (2017). 3D printers as sociable technologies: Taking appropriation infrastructures to the internet of things. ACM Transactions on Computer-Human Interaction, 24(2), 1–28.

Mackay, W. E. (1990). Users and customizable software: A co-adaptive phenomenon. Doctoral dissertation, Massachusetts Institute of Technology.

Malaurent, J., & Karanasios, S. (2020). Learning from workaround practices: The challenge of enterprise system implementations in multinational corporations. Information Systems Journal, 30(4), 639–663.

Marx K. (1976). Capital: A critique of political economy, volume one. Penguin Books, London.

McGrenere, J., & Ho, W. (2000). Affordances: Clarifying and evolving a concept. Proceedings of the Graphics Interface Conference, 179–186.

Meier, F. M., Schmidt, A. L., & Bogers, T. (2021). "They each have their forte": An exploratory diary study of temporary switching behavior between mobile messenger services. Proceedings of iConference, 268–286.

Mekler, E. D., & Hornbæk, K. (2019). A framework for the experience of meaning in human-computer interaction. Proceedings of the SIGCHI Conference on Human Factors in Computing Systems, Article 225 (1–15).

Meschtscherjakov, A., Wilfinger, D., & Tscheligi, M. (2014). Mobile attachment causes and consequences for emotional bonding with mobile phones. Proceedings of the SIGCHI Conference on Human Factors in Computing Systems, 2317–2326.

Mi, X., Qian, F., Zhang, Y., & Wang, X. (2017). An empirical characterization of IFTTT: Ecosystem, usage, and performance. Proceedings of the Internet Measurement Conference, 398–404.

Miller, C. R. (2015). Genre as social action (1984), revisited 30 years later (2014). Letras & Letras, 31(3), 56–72.

Miller, C R. (1994). Rhetorical community: The cultural basis of genre. In A. Freedman & P. Medway (Eds.), Genre and the new rhetoric, 67–78. Taylor & Francis, London.

Miller, C. R. (1984). Genre as social action. Quarterly Journal of Speech, 70(2), 151–167.

Miller, C. R., & Shepherd, D. (2009). Questions for genre theory from the blogosphere. In J. Giltrow & D. Stein (Eds.), Genres in the internet: Issues in the theory of genre, 263–290. John Benjamins, Amsterdam.

Miller, T. (2019). Explanation in artificial intelligence: Insights from the social sciences. Artificial Intelligence, 267, 1–38.

Mørch, A. I. (2011). Evolutionary application development: Tools to make tools and boundary crossing. In H. Isomäki & S. Pekkola (Eds.), Reframing humans in information systems development, 151–171. Springer, London.

Mørch, A. I. (1997). Three levels of end-user tailoring: Customization, integration, and extension. In M. Kyng & L. Mathiassen (Eds.), Computers and design in context, 51–76. MIT Press, Cambridge.

Mørch, A. I., Caruso, V., & Hartley, M. D. (2017). End-user development and learning in Second Life: The evolving artifacts framework with application. In F. Paternò & V. Wulf (Eds.), New perspectives in end-user development, 333–358. Springer, Cham.

Mugge, R., Schoormans, J. P., & Schifferstein, H. N. (2009). Emotional bonding with personalised products. Journal of Engineering Design, 20(5), 467–476.

Nardi, B. (1996). Context and consciousness: Activity theory and human-computer interaction. MIT Press, Cambridge.

Nardi, B. (1993). A small matter of programming: Perspectives on end user computing. MIT Press, Cambridge.

Norman, D. A. (1988): The psychology of everyday things. Basic Books, New York.

Nouwens, M., Griggio, C. F., & Mackay, W. E., (2017). WhatsApp is for family; messenger is for friends: Communication places in app ecosystems. Proceedings of the SIGCHI Conference on Human Factors in Computing Systems, 727–735.

Ochanine, D. A. (1977). Concept of operative image in engineering and general psychology. In B. F. Lomov, V. F. Rubakhin, & V. F. Venda (Eds.), Engineering psychology, 134–149. Science publishers, Moscow.

Odom, W., Pierce, J., Stolterman, E., & Blevis, E. (2009). Understanding why we preserve some things and discard others in the context of interaction design. Proceedings of the SIGCHI Conference on Human Factors in Computing Systems, 1053–1062.

Ollman, B. (1971). Alienation: Marx's conception of man in capitalist society. Cambridge University Press, Cambridge.

Orlikowski, W. J. (2000). Using technology and constituting structures: A practice lens for studying technology in organizations. Organization Science, 11(4), 404–428.

Orlikowski, W. J (1992). The duality of technology: Rethinking the concept of technology in organizations. Organization Science 3(3), 398–427.

Orth, D., Thurgood, C., & Hoven, E.V.D. (2019). Designing meaningful products in the digital age: How users value their technological possessions. ACM Transactions on Computer-Human Interaction, 26(5), 1–28.

Oulasvirta, A., & Blom, J. (2008). Motivations in personalisation behaviour. Interacting with Computers, 20(1), 1–16.

Pastré, P., Mayen, P., & Vergnaud, G. (2006). La didactique professionnelle. Revue française de pédagogie, 154, 145–198 (in French).

Paternò, F., & Santoro, C. (2019). End-user development for personalizing applications, things, and robots. International Journal of Human-Computer Studies, 131, 120–130.

Piaget, J. (1952). The origins of intelligence in children. International Universities Press, New York.

Pierce, J. L., Kostova, T., & Dirks, K. T. (2003). The state of psychological ownership: Integrating and extending a century of research. Review of General Psychology, 7(1), 84–107.

Pilke E. M. (2004). Flow experiences in information technology use. International Journal of Human-Computer Studies, 61(3), 347–357.

Pipek, V., & Wulf, V. (2009). Infrastructuring: Toward an integrated perspective on the design and use of information technology. Journal of the Association for Information Systems, 10(5), 447–473.

Pizza, S., Brown, B., McMillan, D., & Lampinen, A. (2016). Smartwatch in vivo. Proceedings of the SIGCHI Conference on Human Factors in Computing Systems, 5456–5469.

Polites, G. L., & Karahanna, E. (2012). Shackled to the status quo: The inhibiting effects of incumbent system habit, switching costs, and inertia on new system acceptance. MIS Quarterly, 36(1), 21–42.

Poole, M. S., & de Sanctis, G. (1989). Use of group decision support systems as an appropriation process. Proceedings of the Hawaii International Conference on System Sciences, volume 4, 149–150.

Porcheron, M., Fischer, J. E., Reeves, S., & Sharples, S. (2018). Voice interfaces in everyday life. Proceedings of the SIGCHI Conference on Human Factors in Computing Systems, Article 640 (1–12).

Pu, P., Chen, L., & Hu, R. (2012). Evaluating recommender systems from the user's perspective: Survey of the state of the art. User Modeling and User-Adapted Interaction, 22(4–5), 317–355.

Pucihar, K. C., Kljun, M., Mariani, J., & Dix, A. J. (2016). An empirical study of long-term personal project information management. Aslib Journal of Information Management, 68(4), 495–522.

Purington, A., Taft, J. G., Sannon, S., Bazarova, N. N., & Taylor, S. H. (2017). "Alexa is my new BFF": Social roles, user satisfaction, and personification of the Amazon Echo. Proceedings of the SIGCHI Conference on Human Factors in Computing Systems Extended Abstracts, 2853–2859.

Pyszczynski, T., Greenberg J., Koole, S., & Solomon, S. (2010). Experimental existential psychology: Coping with the facts of life. In S. T. Fiske, D. T. Gilbert & G. Lindzey (Eds.), Handbook of social psychology, 724–757. John Wiley & Sons, New York.

Pyszczynski, T., Solomon, S., & Greenberg, J. (2015a). Thirty years of terror management theory: From genesis to revelation. In Olson J. M. & Zanna, P. (Eds.), Advances in experimental social psychology, 1–70. Academic Press, Cambridge.

Pyszczynski, T., Sullivan, D., & Greenberg, J. (2015b). Experimental existential psychology: Living in the shadow of the facts of life. In M. Mikulincer & P. R. Shaver (Eds.), American Psychological Association handbook of personality and social psychology, 297–308. American Psychological Association, Washington DC.

Rabardel, P. (2003). From artefact to instrument. Interacting with Computers, 15(5), 641–645.

Rabardel, P., (2001). Instrument mediated activity in situations. In A. Blandford, J. Vanderdonckt & P. Gray (Eds.), People and computers XV—Interactions without frontiers, 17–30. Springer-Verlag, London.

Rabardel, P., & Pastré, P. (2005). Modèles du sujet pour la conception. Octares Éditions, Toulouse (in French).

Rabassa, V., Sabri, O., & Spaletta, C. (2022). Conversational commerce: Do biased choices offered by voice assistants' technology constrain its appropriation? Technological Forecasting and Social Change, 174, 121292.

Reinke, K., & Chamorro-Premuzic, T. (2014). When email use gets out of control. Computers in Human Behavior, 36(C), 502–509.

Retore, A. P., & Almeida, L.D.A. (2019). Understanding appropriation through end-user tailoring in communication systems: A case study on Slack and WhatsApp. Proceedings of the International Conference on Human-Computer Interaction, 245–264.

Reyes, P., & Tchounikine, P. (2003). Supporting emergence of threaded learning conversations through augmenting interactional and sequential coherence. In B. Wasson, S. Ludvigsen & U. Hoppe (Eds.), Designing for change in networked learning environments, 83–92. Springer, Dordrecht.

Rochat, N., Seifert, L., Guignard, B., & Hauw, D. (2019). An enactive approach to appropriation in the instrumented activity of trail running. Cognitive Processing, 20(4), 459–477.

Rogers, R., & Dunlow, L. (2019). Testing the difference between appearance and ability customization. Communication Design Quarterly Review, 7(2), 7–16.

Rosenberg, H., & Asterhan, C. S. (2018). "WhatsApp, teacher?"—Student perspectives on teacher-student WhatsApp interactions in secondary schools. Journal of Information Technology Education: Research, 17, 205–226.

Russell, D. R. (2009). Uses of activity theory in written communication research. In A. Sannino, H. Daniels & K. D. Gutiérrez (Eds.), Learning and expanding with activity theory, 40–52. Cambridge University Press, New York.

Salovaara, A., Helfenstein, S., & Oulasvirta, A. (2011). Everyday appropriations of information technology: A study of creative uses of digital cameras. Journal of the American Society for Information Science and Technology, 62(12), 2347–2363.

Sappelli, M., Pasi, G. Verberne, S., de Boer, M., & Kraaij, W. (2016). Assessing e-mail intent and tasks in e-mail messages. Information Sciences, 358–359, 1–17.

Sartre J. P. (1946). L'existentialisme est un humanisme. Nagel, Paris (Translation: Existentialism is a Humanism [2007]. Yale University Press, New Haven).

Satchell, C., & Dourish, P. (2009). Beyond the user: Use and non-use in HCI. Proceedings of the Conference of the Australian Computer-Human Interaction Special Interest Group, 9–16.

Schank, R. C. (1999). Dynamic memory revisited. Cambridge University Press, Cambridge.

Schank, R. C., & Abelson, R. P. (1977). Scripts, plans, goals and understanding. Erlbaum, Hillsdale.

Schmitz, K. W., Teng, J. T., & Webb, K. J. (2016). Capturing the complexity of malleable IT use: Adaptive structuration theory for individuals. Mis Quarterly, 40(3), 663–686.

Schryer, C. F., Afros, E., Mian, M., Spafford, M., & Lingard, L. (2009). The trial of the expert witness: Negotiating credibility in child abuse correspondence. Written Communication, 26(3), 215–246.

Schwartz, T., Stevens, G., Jakobi, T., Denef, S., Ramirez, L., Wulf, V., & Randall, D. (2015). What people do with consumption feedback: A long-term living lab study of a home energy management system. Interacting with Computers, 27(6), 551–576.

Sengers, P., Boehner, K., David, S., & Kaye, J. (2005). Reflective design. Proceedings of the Decennial Conference on Critical Computing: Between Sense and Sensibility, 49–53.

Sengers, P., & Gaver, B. (2006). Staying open to interpretation: Engaging multiple meanings in design and evaluation. Proceedings of the ACM Conference on Designing Interactive systems, 99–108.

Sevtsuk, A., Chancey, B., Basu, R., & Mazzarello, M. (2022). Spatial structure of workplace and communication between colleagues: A study of e-mail exchange and spatial relatedness on the MIT campus. Social Networks, 70, 295–305.

Shi, X., Carliner, S., & Wan, W. (2020). Internet-mediated genre studies: An integrative literature review (2005–2019). IEEE Transactions on Professional Communication, 63(4), 279–295.

Siang, T. Y. (2020). What is interaction design? https://www.interaction-design.org/literature/article/what-is-interaction-design.

Singh, N., Tomitsch, M., & Maher, M. L. (2013). Understanding the management and need for awareness of temporal information in email. Proceedings of the Australasian User Interface Conference, 43–51.

Slack (2020). https://slack.com.

Sobreira, P., & Tchounikine, P. (2015). Table-based representations can be used to offer easy-to-use, flexible, and adaptable learning scenario editors. Computers & Education, 80, 15–27.

Sobreira, P., & Tchounikine, P. (2012). A model for flexibly editing CSCL scripts. International Journal of Computer-Supported Collaborative Learning, 7(4), 567–592.

Spinuzzi, C. (2020). "Trying to predict the future": Third-generation activity theory's codesign orientation. Mind, Culture, and Activity, 27(1), 4–18.

Spinuzzi, C. (2003). Tracing genres through organizations: A sociocultural approach to information design. MIT Press, Cambridge.

Spinuzzi, C. (2001). Software development as mediated activity: Applying three analytical frameworks for studying compound mediation. Proceedings of the International Conference on Computer Documentation, 58–67.

Spinuzzi, C., & Zachry, M. (2000). Genre ecologies: An open-system approach to understanding and constructing documentation. Journal of Computer Documentation, 24(3), 169–181.

Stevens, G., & Pipek, V. (2018). Making use: Understanding, studying, and supporting appropriation. In V. Wulf, V. Pipeck, D. Randall, M. Rohde, K. Schmidt & G. Stevens (Eds.), Socio-informatics, a practice-based perspective on the design and use of IT artifacts, 139–176. Oxford University Press, Oxford.

Stevens, G., Pipek, V., & Wulf, V. (2010). Appropriation infrastructure: Mediating appropriation and production work. Journal of Organizational and End User Computing, 22(2), 58–81.

Stevens, G., Quaisser, G., & Klann, M. (2006). Breaking it up: An industrial case study of component-based tailorable software design. In H. Lieberman, F., Paternò, & V. Wulf (Eds.), End User Development, 269–294. Springer, Dordrecht.

Stray, V., & Moe, N. B. (2020). Understanding coordination in global software engineering: A mixed-methods study on the use of meetings and Slack. Journal of Systems and Software, 170, 110717.

Suchman, L. (1987). Plans and situated actions. The problem of human-machine communication. Cambridge University Press, Cambridge.

Suddaby, R. (2010). Construct clarity in theories of management and organizations. Academy of Management Review, 35(3), 346–357.

Sun, Y., Guo, Y., & Zhao, Y. (2020). Understanding the determinants of learner engagement in MOOCs: An adaptive structuration perspective. Computers & Education, 157, 103953.

Sun, Y., Liu, D., Chen, S., Wu, X., Shen, X. L., & Zhang, X (2017). Understanding users' switching behavior of mobile instant messaging applications: An empirical study from the perspective of push-pull-mooring framework. Computers in Human Behavior, 75, 727–738.

Tchounikine, P. (2019a). Framing design for appropriation with zones of proximal evolution Email for PIM. International Journal of Human-Computer Studies, 123, 18–28.

Tchounikine, P. (2019b). Learners' agency and CSCL technologies: Towards an emancipatory perspective. International Journal of Computer-Supported Collaborative Learning, 14(2), 237–250.

Tchounikine, P. (2017). Designing for appropriation: A theoretical account. Human-Computer Interaction, 32(4), 155–195.

Tchounikine, P. (2016). Contribution to a theory of CSCL scripts: Taking into account the appropriation of scripts by learners. International Journal of Computer-Supported Collaborative Learning, 11(3), 349–369.

Tchounikine, P. (2011). Computer science and educational software design—A resource for multidisciplinary work in technology enhanced learning. Springer, Berlin.

Thunderbird (2024). https://www.thunderbird.net.

Tiidenberg, K., & Whelan, A. (2017). Sick bunnies and pocket dumps: "Not-selfies" and the genre of self-representation. Popular Communication, 15(2), 141–153.

Todi, K., Bailly, G., Leiva, L., & Oulasvirta, A. (2021). Adapting user interfaces with model-based reinforcement learning. Proceedings of the SIGCHI Conference on Human Factors in Computing Systems, Article 573 (1–13).

Tractinsky, N. (2018). The usability construct: A dead end? Human-Computer Interaction, 33(2), 131–177.

Tractinsky, N. (2004). Toward the study of aesthetics in information technology. Proceedings of the International Conference on Information Systems, 771–780.

Turner, M., Kitchenham, B., Brereton, P., Charters, S., & Budgen, D. (2010). Does the technology acceptance model predict actual use? A systematic literature review. Information and Software Technology, 52(5), 463–479.

Turner, P. (2005). Affordance as context. Interacting with Computers, 17(6), 787–800.

UML (2017). https://www.omg.org/spec/UML.

Ur, B., Pak Yong Ho, M., Brawner, S., Lee, J., Mennicken, S., Picard, N., Schulze, D. & Littman, M. L. (2016). Trigger-action programming in the wild: An analysis of 200,000 IFTTT recipes. Proceedings of the SIGCHI Conference on Human Factors in Computing Systems, 3227–3231.

Venkatesh, V., Morris, M. G., Davis, G. B., & Davis, F. D. (2003). User acceptance of information technology: Toward a unified view. MIS Quarterly, 27(3), 425–478.

Venkatesh, V., Thong, J. Y., & Xu, X. (2016). Unified theory of acceptance and use of technology: A synthesis and the road ahead. Journal of the association for Information Systems, 17(5), 328–376.

Venkatesh, V., Thong, J. Y., & Xu, X. (2012). Consumer acceptance and use of information technology: Extending the unified theory of acceptance and use of technology. MIS Quarterly, 36(1), 157–178.

Vergnaud, G. (2009). The theory of conceptual fields. Human Development, 52(2), 83–94.

Vergnaud, G. (1998). Towards a cognitive theory of practice. In J. Kilpatrick & A. Sierpinska (Eds.), Mathematics education as a research domain: A search for identity, 227–240. Springer, Dordrecht.

Vyas, D., Chisalita, C. M., & Dix, A. (2017). Organizational affordances: A structuration theory approach to affordances. Interacting with Computers, 29(2), 117–131.

Vygotsky, L. S. (1978). Mind in society: The development of higher psychological processes. Harvard University Press, Cambridge.

Wakkary, R., & Maestri, L. (2007). The resourcefulness of everyday design. Proceedings of the SIGCHI Conference on Creativity & Cognition, 163–172.

Wang, D., Tan, H., & Lu, T. (2017). Why users do not want to write together when they are writing together: Users' rationales for today's collaborative writing practices. Proceedings of the ACM on Human-Computer Interaction, Article 107 (1–18).

Wertsch, J. V. (1998). Mind as action. Oxford University Press, Oxford.

Whitham, R., & Cruickshank, L. (2017). The function and future of the folder. Interacting with Computers, 29(5), 629–647.

Whittaker, S. (2005). Collaborative task management in email. Human-Computer Interaction, 20(1–2), 49–88.

Whittaker, S., Bellotti, V., & Gwizdka, J. (2006). Email in personal information management. Communications of the ACM, 49(1), 68–73.

Wilson, T. D., Reinhard, D. A., Westgate, E. C., Gilbert, D. T., Ellerbeck, N., Hahn, C., Brown, C. L., & Shaked, A. (2014). Just think: The challenges of the disengaged mind. Science, 345(6192), 75–77.

Wiseman, S., & Gould, S. J. (2018). Repurposing emoji for personalised communication: Why [pizza icon] means "I love you." Proceedings of the SIGCHI Conference on Human Factors in Computing Systems, Article 152 (1–10).

Wonderbot (2020). https://wonder-bot.com/slack.

Wright, E. O. (2010). Envisioning real utopias. Verso, London.

Wulf, V., Pipek, V. & Won, M. (2008). Component-based tailorability: Enabling highly flexible software applications. International Journal of Human-Computer Studies, 66(1), 1–22.

Yalom, I. (1980). Existential psychotherapy. Basic Books, New York.

Yan, M., Filieri, R., & Gorton, M. (2021). Continuance intention of online technologies: A systematic literature review. International Journal of Information Management, 58(1), 102315.

Yates, J., & Orlikowski, W. (2007). The PowerPoint presentation and its corollaries: How genres shape communicative action in organizations. In M. Zachry & C. Thralls (Eds.), The cultural turn: Communicative practices in workplaces and the professions, 67–92. Baywood, Amityville.

Yates, J. & Orlikowski, W. (2002). Genre systems: Structuring interaction through communicative norms. Journal of Business Communication, 39(1), 13–35.

Yates, J., Orlikowski, W. J., & Okamura, K. (1999). Explicit and implicit structuring of genres in electronic communication: Reinforcement and change of social interaction. Organization Science, 10(1), 83–103.

Zimmerman, J. (2009). Designing for the self: Making products that help people become the person they desire to be. Proceedings of the SIGCHI Conference on Human Factors in Computing Systems, 395–404.

Index

Publisher contact:
The MIT Press
Massachusetts Institute of Technology
77 Massachusetts Avenue, Cambridge, MA 02139
mitpress.mit.edu

EU Authorised Representative:
Easy Access System Europe, Mustamäe tee 50,
10621 Tallinn, Estonia
gpsr.requests@easproject.com

Printed by Integrated Books International,
United States of America